SpringerBriefs in Optimization

SpringerBriefs in Optimization showcases algorithmic and theoretical techniques, case studies, and applications within the broad-based field of optimization. Manuscripts related to the ever-growing applications of optimization in applied mathematics, engineering, medicine, economics, and other applied sciences are encouraged.

For further volumes:
http://www.springer.com/series/8918

Joël Blot • Naïla Hayek

Infinite-Horizon Optimal Control in the Discrete-Time Framework

Springer

Joël Blot
Université Paris 1 Panthéon-Sorbonne
Paris, France

Naïla Hayek
Université Paris 2 Panthéon-Assas
Paris, France

ISSN 2190-8354 ISSN 2191-575X (electronic)
ISBN 978-1-4614-9037-1 ISBN 978-1-4614-9038-8 (eBook)
DOI 10.1007/978-1-4614-9038-8
Springer New York Heidelberg Dordrecht London

Library of Congress Control Number: 2013950905

Mathematics Subject Classification: 49J21, 65K05, 39A99, 49K99 93C55

Printed on acid-free paper

Springer is part of Springer Science+Business Media (www.springer.com)

Preface

Optimal control theory in infinite horizon and discrete time offers a setting which is used in a lot of problems from various scientific fields: economics, management, sustainable development of fisheries and of forests, biology, and medicine. The theory of differential equations is not well known by all specialists of scientific fields, except mathematicians and physicists, whereas understanding the meaning of the equations of a discrete-time dynamical system does not necessitate sophisticated mathematical tools. So, in our opinion, discrete-time models can simplify the communication between mathematicians and researchers of other scientific fields. This opinion is not against the continuous-time models. When both discrete-time modeling and continuous-time modeling of the same phenomenon exist, the comparison between their respective results can provide interesting consequences.

In finite-horizon continuous-time optimal control theory, two main historical approaches exist: Pontryagin's approach and Bellman's approach. In infinite-horizon discrete-time optimal control problems, the dynamic programming of Bellman is currently used. In this book, we want to present in a mathematically rigorous way a treatment of Pontryagin's viewpoint for infinite-horizon discrete-time optimal control problems.

Pontryagin's viewpoint provides necessary conditions of optimality which are laws that the optimal solutions ought to satisfy, and these laws possess a meaning in the considered phenomenon. Moreover the role of necessary conditions of optimality is to narrow the set of all processes which are candidates to be solutions of the problem, and this can also improve the modeling. In some cases, it is also possible to formulate sufficient conditions of optimality in the spirit of the conditions initiated by Seierstad and Sydsaeter for the continuous-time problems.

In the first two chapters, we use an approach of reduction to finite horizon which consists of associating to an infinite-horizon problem a family of finite-horizon problems. In the third chapter, we use another approach for the bounded-process problems, which is based on nonlinear functional analysis in Banach spaces.

In Chap. 1, we present the problems and we give the tools of the finite-horizon case which can be translated into static optimization. We use various multiplier rules that are presented in Appendix B. In Chap. 2, we give necessary conditions theorems for infinite-horizon optimal control problems, under the form of weak or strong Pontryagin principles. We also study problems under constraints and multiobjective optimal control problems. The special case of the bounded processes is treated in Chap. 3, where necessary conditions and sufficient conditions of optimality are given. In Appendix A, we give some elements concerning sequences and sequence spaces, and in Appendix B, we provide many static optimization theorems which are essential for our approaches.

Paris, France

Joël Blot
Naïla Hayek

Contents

Chapter 1
Presentation of the Problems and Tools of the Finite Horizon

1.1 Introduction

In Sect. 1.2 of this first chapter, we formulate the infinite-horizon discrete-time optimal control problems that we study. The considered systems are governed by difference equations or by difference inequations. We define four optimality criterions on such systems.

In Sect. 1.3, we describe a method that we call the reduction to finite horizon: we associate to an optimal process of an infinite-horizon problem a sequence of finite-horizon problems for which the restrictions of the optimal process are solutions.

In Sect. 1.4, we begin to specify the notion of strong maximum principle and of weak maximum principle and we use the contribution of Boltyanskii to understand interesting differences between them. To obtain weak maximum principles on finite-horizon problems we use the multiplier rules of Halkin and of Clarke. To obtain strong maximum principles on finite-horizon problems we present a result of Michel and we adapt a condition of Pschenichnyi.

1.2 The Problems

1.2.1 The Controlled Dynamical Systems

The discrete time is denoted by the letter $t \in \mathbb{N}$, where $\mathbb{N}$ is the set of the nonnegative integer numbers. For all $t \in \mathbb{N}$, X_t is a nonempty subset of $\mathbb{R}^n$ (n is a fixed positive integer), U_t is a nonempty subset of $\mathbb{R}^d$ (d is a positive integer), and $f_t : X_t \times U_t \to X_{t+1}$ is a function.

The usual order of $\mathbb{R}^n$ is $x = (x^1, x^2, \ldots, x^n) \leq (y^1, y^2, \ldots, y^n) = y$ defined by $x^i \leq y^i$ for all $i \in \{1, \ldots, n\}$. And $x < y$ means that $x \leq y$ and $x \neq y$.

J. Blot and N. Hayek, *Infinite-Horizon Optimal Control in the Discrete-Time Framework*, SpringerBriefs in Optimization, DOI 10.1007/978-1-4614-9038-8_1,

To abridge the writing we use the notation $\underline{x} := (x_t)_{t \in \mathbb{N}} \in \prod_{t \in \mathbb{N}} X_t$ (the cartesian product) and also $\underline{u} := (u_t)_{t \in \mathbb{N}} \in \prod_{t \in \mathbb{N}} U_t$.

We work with two families of controlled dynamical systems: difference equations and difference inequations. They are

(DE) $\quad x_{t+1} = f_t(x_t, u_t)$

and

(DI) $\quad x_{t+1} \leq f_t(x_t, u_t)$.

The variable x_t is called the *state variable* and the variable u_t is called the *control variable*.

When we fix an initial state $\eta \in X_0$, we denote by Adm^e_η the set of all processes $(\underline{x}, \underline{u}) \in \prod_{t \in \mathbb{N}} X_t \times \prod_{t \in \mathbb{N}} U_t$ which satisfy (DE) at each time $t \in \mathbb{N}$ and such that $x_0 = \eta$. These processes are called the *admissible processes* for (DE) and η. The letter e as upper index means *equation* and the lower index indicates the initial value of the state. We denote by Adm^i_η the set of all processes $(\underline{x}, \underline{u}) \in \prod_{t \in \mathbb{N}} X_t \times \prod_{t \in \mathbb{N}} U_t$ which satisfy (DI) at each time $t \in \mathbb{N}$ and such that $x_0 = \eta$. These processes are called the *admissible processes* for (DI) and η. The letter i as upper index means *inequation* and η as lower index indicates the initial state.

1.2.2 The Optimality Criterions

For all $t \in \mathbb{N}$, we consider a function $\phi_t : X_t \times U_t \to \mathbb{R}$.

The first optimality criterion that we consider is defined by using the functional

$$J(\underline{x}, \underline{u}) := \sum_{t=0}^{+\infty} \phi_t(x_t, u_t)$$

when $(\underline{x}, \underline{u}) \in \mathrm{Adm}^a_\eta$ where $a \in \{e, i\}$. Clearly this functional is not always defined. And so we introduce the *domain* of J, denoted by $\mathrm{Dom}^a_\eta(J)$, as the set of all $(\underline{x}, \underline{u}) \in \mathrm{Adm}^a_\eta$ such that the series $\sum_{t=0}^{+\infty} \phi_t(x_t, u_t)$ converges in $\mathbb{R}$ (toward to a finite real number). And so we can define the first problems

$(\mathcal{P}^n_a)$ $\quad$ Maximize $J(\underline{x}, \underline{u})$ when $(\underline{x}, \underline{u}) \in \mathrm{Dom}^a_\eta(J)$.

The upper index n means *natural*; such problems arise so naturally in various theories as we shall see later.

The second optimality criterion, which is stronger than the first one, is defined as follows:

$(\mathcal{P}^s_a)$ $\quad$ Find $(\underline{\hat{x}}, \underline{\hat{u}}) \in \mathrm{Dom}^a_\eta(J)$ such that, for all $(\underline{x}, \underline{u}) \in \mathrm{Adm}^a_\eta$,

$$J(\underline{\hat{x}}, \underline{\hat{u}}) \geq \limsup_{T \to +\infty} \sum_{t=0}^{T} \phi_t(x_t, u_t).$$

The upper index s means *strong*.

In the first two optimality criterions, the optimal value of J is finite, i.e., belongs to $\mathbb{R}$. The following two problems allow to compare admissible processes, when several admissible processes give the value $+\infty$ to the functional J. The following problem defines the so-called overtaking optimality:

($\mathscr{P}_a^o$) Find $(\underline{\hat{x}}, \underline{\hat{u}}) \in \mathrm{Adm}_\eta^a$ such that, for all $(\underline{x}, \underline{u}) \in \mathrm{Adm}_\eta^a$,

$$\liminf_{T \to +\infty} \sum_{t=0}^{T} (\phi_t(\hat{x}_t, \hat{u}_t) - \phi_t(x_t, u_t)) \geq 0.$$

The following problem defines the so-called weak overtaking optimality:

($\mathscr{P}_a^w$) Find $(\underline{\hat{x}}, \underline{\hat{u}}) \in \mathrm{Adm}_\eta^a$ such that, for all $(\underline{x}, \underline{u}) \in \mathrm{Adm}_\eta^a$,

$$\limsup_{T \to +\infty} \sum_{t=0}^{T} (\phi_t(\hat{x}_t, \hat{u}_t) - \phi_t(x_t, u_t)) \geq 0.$$

Remark 1.1. The criterion of ($\mathscr{P}_a^n$) is a functional in the form

$$J(\underline{x}, \underline{u}) = \lim_{T \to +\infty} J_T(\underline{x}, \underline{u}),$$

where $J_T(\underline{x}, \underline{u}) := \sum_{t=0}^{T} \phi_t(x_t, u_t)$, i.e., a criterion which is a limit of a sequence of functionals. The criterion of ($\mathscr{P}_a^s$) is a functional in the form

$$(\underline{x}, \underline{u}) \mapsto \limsup_{T \to +\infty} J_T(\underline{x}, \underline{u})$$

with the additional condition on the maximizer: $(\underline{\hat{x}}, \underline{\hat{u}}) \in \mathrm{Dom}_\eta^a(J)$. Note that the last two optimality criterions are not in the form of the maximization of a functional. As it is written in [7], we can note that the binary relation $\mathscr{R}$ on Adm_η^a defined by

$$(\underline{x}, \underline{u}) \mathscr{R} (\underline{y}, \underline{v}) \Longleftrightarrow \liminf_{T \to +\infty} \sum_{t=0}^{T} (\phi_t(y_t, v_t) - \phi_t(x_t, u_t)) \geq 0$$

is a pre-ordering (reflexive and transitive). And so to solve ($\mathscr{P}_a^o$) is to find a maximum for a pre-ordering.

The "natural" criterion is used by Ramsey in [78] (published in 1928) and by Hotelling in [53] (published in 1931) in the continuous-time framework. The overtaking optimality criterion is used by von Weizsäcker in [90] (published in 1965) in the continuous-time framework; it is used by Brock in [33] (published in 1970) in the discrete-time framework in the spirit of a previous work of Gale [45] (published in 1967). Note that Brock does not cite the work of von Weizsäcker [90].

Remark 1.2. There exist other notions of optimality for such dynamical systems. If we translate the minimization problem treated by Flam and Wets in [42, 44] into a maximization problem, the optimality criterion is the maximization of the functional $(\underline{x}, \underline{u}) \mapsto \limsup_{T\to+\infty} J_T(\underline{x}, \underline{u})$. In [72] Michel attributes to Gale our notion of overtaking optimality (Definition 2, p. 707) and he calls it the *catching up optimality*. Michel attributes to Brock the following notion (Definition 3, p. 707): $(\hat{\underline{x}}, \hat{\underline{u}})$ is optimal when $\liminf_{T\to+\infty} \sum_{t=0}^{T} (\phi_t(\hat{x}_t, \hat{u}_t) - \phi_t(x_t, u_t)) > 0$ for all $(\underline{x}, \underline{u}) \in \mathrm{Adm}_\eta^a \setminus \{(\hat{\underline{x}}, \hat{\underline{u}})\}$; he calls it the overtaking optimality. He uses the term of weak overtaking optimality as we use it.

The following result gives comparisons between the considered problems.

Proposition 1.1. *The following assertions hold.*

(i) If $(\hat{\underline{x}}, \hat{\underline{u}})$ is a solution of $(\mathscr{P}_a^o)$, then $(\hat{\underline{x}}, \hat{\underline{u}})$ is a solution of $(\mathscr{P}_a^w)$.
(ii) If $(\hat{\underline{x}}, \hat{\underline{u}})$ is a solution of $(\mathscr{P}_a^s)$, then $(\hat{\underline{x}}, \hat{\underline{u}})$ is a solution of $(\mathscr{P}_a^n)$.
(iii) If $(\hat{\underline{x}}, \hat{\underline{u}}) \in \mathrm{Dom}_\eta^a(J)$ and if $(\hat{\underline{x}}, \hat{\underline{u}})$ is a solution of $(\mathscr{P}_a^o)$, then $(\hat{\underline{x}}, \hat{\underline{u}})$ is a solution of $(\mathscr{P}_a^s)$.

Proof. Since the limsup of a real sequence is always greater than the liminf of the same sequence, we obtain (i). Since the limit of a convergent sequence is equal to its limsup, we obtain (ii). To prove (iii), for all $(\underline{x}, \underline{u}) \in \mathrm{Adm}_\eta^a$, we have

$$\begin{aligned} 0 &\le \liminf_{T\to+\infty} \sum_{t=0}^{T} (\phi_t(\hat{x}_t, \hat{u}_t) - \phi_t(x_t, u_t)) \\ &= \lim_{T\to+\infty} \sum_{t=0}^{T} \phi_t(\hat{x}_t, \hat{u}_t) + \liminf_{T\to+\infty} \sum_{t=0}^{T} (-\phi_t(x_t, u_t)) \\ &= J(\hat{\underline{x}}, \hat{\underline{u}}) - \limsup_{T\to+\infty} \sum_{t=0}^{T} \phi_t(x_t, u_t), \end{aligned}$$

and so $(\hat{\underline{x}}, \hat{\underline{u}})$ is a solution of $(\mathscr{P}_a^s)$. □

For the properties of limsup and liminf used in the previous proof, see Appendix A. Note also that we do not need any special assumption to obtain the previous proposition.

We will use the *Hamiltonian of Pontryagin* (also called pre-Hamiltonian) to treat these problems. It is the function $H_t : X_t \times U_t \times \mathbb{R}^{n*} \times \mathbb{R} \to \mathbb{R}$, for all $t \in \mathbb{N}$, where $\mathbb{R}^{n*} = \mathscr{L}(\mathbb{R}^n, \mathbb{R})$ the dual space of $\mathbb{R}^n$, defined by

$$H_t(x_t, u_t, p, \lambda_0) := \lambda_0 \phi_t(x_t, u_t) + \langle p, f_t(x_t, u_t) \rangle \tag{1.1}$$

where $\langle .,. \rangle$ is the duality bracket, $\langle .,. \rangle : \mathbb{R}^{n*} \times \mathbb{R}^n \to \mathbb{R}$, $\langle p, x \rangle := p(x)$.

1.2.3 Motivations

A historical motivation for infinite-horizon variational problems and infinite-horizon optimal control problems is found in the macroeconomic optimal growth theory in the works of Ramsey, [78], and Hotelling, [53]. In such a theory, an agent represents itself and all its progeny, and the infinite horizon avoids to deal with the problems of the end of the world. A part of this literature uses the continuous time, and another part uses the discrete time. There are numerous works on this subject: [4, 9, 25, 35, 36, 68, 80, 83, 87, 91]. Among recent books which use the discrete time, we can cite [61, 65, 86].

Another important field of knowledge which uses the infinite-horizon optimal control is the management of natural resources as forests and fisheries, [37]. More generally, the study of some aspect of sustainable development naturally leads to a framework where a final time does not exist.

We can find further motivations issued from Physics in [93].

1.2.4 Comparisons Between Problems on (DE) and Problems on (DI)

To establish results that compare optimality for problems on (DE) with optimality for problems on (DI) we introduce the following list of conditions in which the letters CA mean "comparison assumptions":

(CA 1) For all $t \in \mathbb{N}$, for all $u_t \in U_t$, $\phi_t(., u_t)$ is increasing.
(CA 2) For all $t \in \mathbb{N}$, for all $u_t \in U_t$, $f_t(., u_t)$ is increasing.
(CA 3) For all $t \in \mathbb{N}$, ϕ_t is nonnegative.
(CA 4) For all $t \in \mathbb{N}$, for all $(y_{t+1}, y_t, u_t) \in X_{t+1} \times X_t \times U_t$ such that $y_{t+1} \leq f_t(y_t, u_t)$, there exists $v_t \in U_t$ such that $v_t \geq u_t$ and such that $y_{t+1} = f_t(y_t, v_t)$.
(CA 5) For all $t \in \mathbb{N}$, for all $x_t \in X_t$, $\phi_t(x_t, .)$ is increasing.

When $(\underline{\hat{x}}, \underline{\hat{u}}) \in \mathrm{Adm}^e_\eta$, we consider the condition

(CA, $(\underline{\hat{x}}, \underline{\hat{u}})$) For all $t \in \mathbb{N}$, for all $z_t \in X_t$, there exists $s \in \mathbb{N}_*$ and there exists $(v_t, \ldots, v_{t+s-1}) \in \prod_{j=0}^{s-1} U_{t+j}$ such that by setting $z_{t+j+1} := f_{t+j}(z_{t+j}, v_{t+j})$ for $j \in \{0, \ldots, s-1\}$, we have $z_{t+s} = \hat{x}_{t+s}$.

Using the vocabulary of the control theory, this last condition can be viewed as a condition of reachability in finite time. In the special case where f_t, X_t, U_t are independent of t, the notion of controllability in finite time as defined in [84] (p. 81) or in [63] (p. 98) implies the condition (CA, $(\underline{\hat{x}}, \underline{\hat{u}})$).

In a first time, we provide relations for the overtaking optimality, between the solutions of $(\mathscr{P}^o_e)$ and the solutions of $(\mathscr{P}^o_i)$.

Theorem 1.1. *Under (CA 1) and (CA 2), if $(\underline{\hat{x}}, \underline{\hat{u}})$ is a solution of $(\mathscr{P}^o_e)$, then it is also a solution of $(\mathscr{P}^o_i)$.*

Proof. Let $(\underline{y}, \underline{u}) \in \text{Adm}^i_\eta$. If $(\underline{y}, \underline{u}) \notin \text{Adm}^e_\eta$ we set

$$T := \min\{t \in \mathbb{N}_* \, : \, y_{t+1} < f_t(y_t, u_t)\}.$$

We set $x_t := y_t$ when $t \leq T$, $x_{T+1} := f_T(x_T, u_T)$ and by induction we define $x_{t+1} := f_t(x_t, u_t)$ when $t > T + 1$. And so we have built $(\underline{x}, \underline{u}) \in \text{Adm}^e_\eta$. Now using (CA 2) we note that we have

$$x_{t+1} = f_t(x_t, u_t) \geq f_t(y_t, u_t) \geq y_{t+1},$$

from which we deduce by induction that $x_t \geq y_t$ for all $t \in \mathbb{N}$. Then using (CA 1) we obtain that $\phi_t(x_t, u_t) \geq \phi_t(y_t, u_t)$, for all $t \in \mathbb{N}$, which implies $\liminf_{s \to +\infty} \sum_{t=0}^{s} (\phi_t(x_t, u_t) - \phi_t(y_t, u_t)) \geq 0$.

And so we have proven that for all $(\underline{y}, \underline{u}) \in \text{Adm}^i_\eta$ there exists $(\underline{x}, \underline{u}) \in \text{Adm}^e_\eta$ such that $\liminf_{s \to +\infty} \sum_{t=0}^{s} (\phi_t(x_t, u_t) - \phi_t(y_t, u_t)) \geq 0$; when $(\underline{y}, \underline{u}) \in \text{Adm}^e_\eta$ it suffices to take $\underline{x} := \underline{y}$.

Then since we have $\liminf_{s \to +\infty} \sum_{t=0}^{s} (\phi_t(\hat{x}_t, \hat{u}_t) - \phi_t(x_t, u_t)) \geq 0$, using the transitivity of the binary relation $\mathcal{R}$ in Remark 1.1, we obtain $\liminf_{s \to +\infty} \sum_{t=0}^{s} (\phi_t(\hat{x}_t, \hat{u}_t) - \phi_t(y_t, u_t)) \geq 0$; and so $(\underline{\hat{x}}, \underline{\hat{u}})$ is a solution of $(\mathcal{P}^o_i)$. □

Theorem 1.2. *Under (CA 4) and (CA 5), if $(\underline{\hat{x}}, \underline{\hat{u}})$ is a solution of $(\mathcal{P}^o_e)$, then it is also a solution of $(\mathcal{P}^o_i)$.*

Proof. Using (CA 4), when $(\underline{y}, \underline{u}) \in \text{Adm}^i_\eta$ there exists $\underline{v} \in \prod_{t \in \mathbb{N}} U_t$ such that $(\underline{y}, \underline{v}) \in \text{Adm}^e_\eta$ and such that $v_t \geq u_t$ for all $t \in \mathbb{N}$. Using (CA 5) we deduce that $\phi_t(y_t, v_t) \geq \phi_t(y_t, u_t)$ for all $t \in \mathbb{N}$. Then we have $\liminf_{T \to +\infty} \sum_{t=0}^{T} (\phi_t(y_t, v_t) - \phi_t(y_t, u_t)) \geq 0$. We know that $\liminf_{T \to +\infty} \sum_{t=0}^{T} (\phi_t(\hat{x}_t, \hat{u}_t) - \phi_t(y_t, v_t)) \geq 0$, and then using the transitivity of the binary relation $\mathcal{R}$ in Remark 1.1, we obtain $\liminf_{T \to +\infty} \sum_{t=0}^{T} (\phi_t(\hat{x}_t, \hat{u}_t) - \phi_t(y_t, u_t)) \geq 0$ that finishes the proof. □

To prove the following theorems we will need the following lemma.

Lemma 1.1. *Under (CA 3), let $(\underline{\hat{x}}, \underline{\hat{u}})$ be a solution of $(\mathcal{P}^n_e)$ such that the condition (CA, $(\underline{\hat{x}}, \underline{\hat{u}})$) holds. Then* $\text{Adm}^e_\eta = \text{Dom}^e_\eta(J)$.

Proof. Since $\text{Dom}^e_\eta(J) \subset \text{Adm}^e_\eta$ it suffices to prove the inverse inclusion. We proceed by contradiction, we consider $(\underline{x}, \underline{u}) \in \text{Adm}^e_\eta$ such that $(\underline{x}, \underline{u}) \notin Dom^e_\eta(J)$. Since $\phi_t(x_t, u_t) \geq 0$ for all $t \in \mathbb{N}$, the only possibility for not having the convergence of the series in $\mathbb{R}$ is $\sum_{t=0}^{+\infty} \phi_t(x_t, u_t) = +\infty$. Since $J(\underline{\hat{x}}, \underline{\hat{u}}) < +\infty$, there exists $T \in \mathbb{N}_*$ such that $\sum_{t=0}^{T} \phi_t(x_t, u_t) > J(\underline{\hat{x}}, \underline{\hat{u}})$. Using the condition (CA, $(\underline{\hat{x}}, \underline{\hat{u}})$) with $t = T$ and $z_t = x_t$ we build

$$y_t := \begin{cases} x_t \text{ if } t \leq T \\ z_t \text{ if } T+1 \leq t \leq T+1+s \\ \hat{x}_t \text{ if } t \geq T+1+s \end{cases}$$

and

$$w_t := \begin{cases} u_t \text{ if } t \leq T \\ v_t \text{ if } T+1 \leq t \leq T+1+s \\ \hat{u}_t \text{ if } t \geq T+1+s \end{cases}$$

where s, z_t, v_t are provided by (CA, $(\underline{\hat{x}}, \underline{\hat{u}})$). Note that $y_t \in X_t$ since $x_t, z_t, \hat{x}_t \in X_t$, and $w_t \in U_t$ since $u_t, v_t, \hat{u}_t \in U_t$ for all $t \in \mathbb{N}$. Moreover, when $t < T$ we have $y_{t+1} = x_{t+1} = f_t(x_t, u_t) = f_t(y_t, u_t)$, when $T \leq t \leq T+s$ we have $y_{t+1} = z_{t+1} = f_t(z_t, v_t) = f_t(y_t, w_t)$ after (CA, $(\underline{\hat{x}}, \underline{\hat{u}})$), and when $t \geq T+1+s$ we have $y_{t+1} = \hat{x}_{t+1} = f_t(\hat{x}_t, \hat{u}_t) = f_t(y_t, w_t)$. And so we obtain that $(\underline{y}, \underline{w}) \in \text{Adm}^e_\eta$. From the inequality $\sum_{t=T+1+s}^{+\infty} \phi_t(y_t, w_t) = \sum_{t=T+1+s}^{+\infty} \phi_t(\hat{x}_t, \hat{u}_t) < +\infty$, we deduce that $(\underline{y}, \underline{w}) \in \text{Dom}^e_\eta(J)$. Using (CA 3), we have

$$J(\underline{y}, \underline{w}) \geq \sum_{t=0}^{T} \phi_t(y_t, w_t) = \sum_{t=0}^{T} \phi_t(x_t, u_t) > J(\underline{\hat{x}}, \underline{\hat{u}}),$$

that is impossible since $(\underline{\hat{x}}, \underline{\hat{u}})$ is optimal. And so the proof is complete. □

The following theorems provide relations between the solutions of $(\mathcal{P}^n_e)$ and those of $(\mathcal{P}^n_i)$.

Theorem 1.3. *Under (CA 1), (CA 2), and (CA 3), if $(\underline{\hat{x}}, \underline{\hat{u}})$ is a solution of $(\mathcal{P}^n_e)$ for which the condition (CA, $(\underline{\hat{x}}, \underline{\hat{u}})$) holds, then it is also a solution of $(\mathcal{P}^n_i)$.*

Proof. When $(\underline{y}, \underline{v}) \in \text{Dom}^i_\eta(J) \subset \text{Adm}^i_\eta$, we build $\underline{x}$ by setting $x_0 := \eta$ and by induction $x_{t+1} = f_t(x_t, v_t)$ for all $t \in \mathbb{N}_*$. Note that we have $(\underline{x}, \underline{v}) \in \text{Adm}^e_\eta$. After (CA 2), since $x_0 = y_0 = \eta$ we have $y_1 \leq f_0(y_0, v_0) = f_0(x_0, v_0) = x_1$, and working by induction we obtain $y_t \leq x_t$ for all $t \in \mathbb{N}$. Therefore, using (CA 1), we deduce $\phi_t(y_t, v_t) \leq \phi_t(x_t, v_t)$ for all $t \in \mathbb{N}$, and using (CA 3) to ensure the convergence of the series in $[-\infty, +\infty]$, we obtain $\sum_{t=0}^{+\infty} \phi_t(y_t, v_t) \leq \sum_{t=0}^{+\infty} \phi_t(x_t, v_t)$. But using Lemma 1.1 we know that $J(\underline{x}, \underline{v}) < +\infty$, and moreover we have $J(\underline{y}, \underline{v}) \leq J(\underline{x}, \underline{v}) \leq J(\underline{\hat{x}}, \underline{\hat{u}})$. □

Theorem 1.4. *Under (CA 3), (CA 4), and (CA 5), if $(\underline{\hat{x}}, \underline{\hat{u}})$ is a solution of $(\mathcal{P}^n_e)$ for which the condition (CA, $(\underline{\hat{x}}, \underline{\hat{u}})$) holds, then it is also a solution of $(\mathcal{P}^n_i)$.*

Proof. When $(\underline{y}, \underline{u}) \in \text{Dom}^i_\eta(J)$, using (CA 4) we build $\underline{v} \in \prod_{t \in \mathbb{N}} U_t$ such that $(\underline{y}, \underline{v}) \in \text{Adm}^e_\eta$ which satisfies $v_t \geq u_t$ for all $t \in \mathbb{N}$. Then Lemma 1.1 allows to say that $(\underline{y}, \underline{v}) \in \text{Dom}^e_\eta(J)$. Using (CA 5) we obtain $\phi_t(y_t, v_t) \geq \phi_t(y_t, u_t)$, for all $t \in \mathbb{N}$, which implies $J(\underline{y}, \underline{u}) \leq (\underline{y}, \underline{v}) \leq J(\underline{\hat{x}}, \underline{\hat{u}})$. □

The results of this subsection come from the paper of Blot [11].

1.3 The Reduction to the Finite Horizon

The **general principle** is the following one: when a process is optimal on $\mathbb{N}$ (until infinity), then for all $T \in \mathbb{N}_*$ its restriction to $[0, T] \cap \mathbb{N}$ is optimal by fixing the final condition at T.

Note that this is different from the principle of Dynamic Programming of Bellman which is when a process is optimal on $\mathbb{N}$ (until infinity), then for all $T \in \mathbb{N}_*$ its restriction to $[T, +\infty) \cap \mathbb{N}$ is optimal by fixing the initial condition at T. We give a rigorous statement of this general principle.

Proposition 1.2. *Let* $(\underline{\hat{x}}, \underline{\hat{u}})$ *be a solution of* $(\mathscr{P}_e^n)$, $(\mathscr{P}_e^s)$, $(\mathscr{P}_e^o)$, *or* $(\mathscr{P}_e^w)$ *(respectively* $(\mathscr{P}_i^n)$, $(\mathscr{P}_i^s)$, $(\mathscr{P}_i^o)$, *or* $(\mathscr{P}_i^w)$*) and let* $T \in \mathbb{N}_*$. *Then the restriction* $((\hat{x}_0, \ldots, \hat{x}_T), (\hat{u}_0, \ldots, \hat{u}_{T-1}))$ *is an optimal solution of the following finite-horizon problem:*

$$(\mathscr{F}^e(T, \eta, \hat{x}_T)) \begin{cases} \text{Maximize } F_T((x_t)_{0 \le t \le T}, (u_t)_{0 \le t \le T-1}) := \sum_{t=0}^{T-1} \phi_t(x_t, u_t) \\ \quad \text{when } \forall t \in \{0, \ldots, T\}, x_t \in X_t \\ \qquad \forall t \in \{0, \ldots, T-1\}, u_t \in U_t \\ \qquad \forall t \in \{0, \ldots, T-1\}, x_{t+1} = f_t(x_t, u_t) \\ \qquad x_0 = \eta, x_T = \hat{x}_T \end{cases}$$

(respectively

$$(\mathscr{F}^i(T, \eta, \hat{x}_T)) \begin{cases} \text{Maximize } F_T((x_t)_{0 \le t \le T}, (u_t)_{0 \le t \le T-1}) := \sum_{t=0}^{T-1} \phi_t(x_t, u_t) \\ \quad \text{when } \forall t \in \{0, \ldots, T\}, x_t \in X_t \\ \qquad \forall t \in \{0, \ldots, T-1\}, u_t \in U_t \\ \qquad \forall t \in \{0, \ldots, T-1\}, x_{t+1} \le f_t(x_t, u_t) \\ \qquad x_0 = \eta, x_T = \hat{x}_T) \end{cases}.$$

Proof. We proceed by contradiction. We assume that $((\hat{x}_0, \ldots, \hat{x}_T), (\hat{u}_0, \ldots, \hat{u}_{T-1}))$ is not optimal for $(\mathscr{F}^e(T, \eta, \hat{x}_T))$. Then there exists $((z_0, \ldots, z_T), (w_0, \ldots, w_{T-1}))$ which is admissible for $(\mathscr{F}^e(T, \eta, \hat{x}_T))$ such that

$$F_T((z_t)_{0 \le t \le T}, (w_t)_{0 \le t \le T-1}) > F_T((\hat{x}_t)_{0 \le t \le T}, (\hat{u}_t)_{0 \le t \le T-1}).$$

When $t > T$ we set $z_t := \hat{x}_t$ and when $t > T-1$ we set $w_t := \hat{u}_t$. And so we have $\underline{z} \in \prod_{t \in \mathbb{N}} X_t$ and $\underline{w} \in \prod_{t \in \mathbb{N}} U_t$. When $t \le T-1$ we have $z_{t+1} = f_t(z_t, w_t)$, and when $t \ge T$ we have $z_{t+1} = \hat{x}_{t+1} = f_t(\hat{x}_t, \hat{u}_t) = f_t(z_t, w_t)$, and so $(\underline{z}, \underline{w}) \in \mathrm{Adm}_\eta^e$.

For $(\mathscr{P}_e^n)$, note that $\sum_{t=T+1}^{+\infty} \phi_t(z_t, w_t) = \sum_{t=T+1}^{+\infty} \phi_t(\hat{x}_t, \hat{u}_t) < +\infty$ implies $(\underline{z}, \underline{w}) \in \mathrm{Dom}_\eta^e(J)$, and

$$
\begin{aligned}
J(\underline{z}, \underline{w}) &= \sum_{t=0}^{T} \phi_t(z_t, w_t) + \sum_{t=T+1}^{+\infty} \phi_t(z_t, w_t) \\
&= \sum_{t=0}^{T} \phi_t(z_t, w_t) + \sum_{t=T+1}^{+\infty} \phi_t(\hat{x}_t, \hat{u}_t) \\
&> \sum_{t=0}^{T} \phi_t(\hat{x}_t, \hat{u}_t) + \sum_{t=T+1}^{+\infty} \phi_t(\hat{x}_t, \hat{u}_t) \\
&= J(\underline{\hat{x}}, \underline{\hat{u}})
\end{aligned}
$$

which is a contradiction.

For $(\mathscr{P}_e^s)$ the reasoning is similar to that of $(\mathscr{P}_e^n)$.

For $(\mathscr{P}_e^o)$, note that, for all $S \in \mathbb{N}$ such that $S > T$, we have

$$
\begin{aligned}
\sum_{t=0}^{S} (\phi_t(z_t, w_t) - \phi_t(\hat{x}_t, \hat{u}_t)) &= \sum_{t=0}^{T} (\phi_t(z_t, w_t) - \phi_t(\hat{x}_t, \hat{u}_t)) \\
&+ \sum_{t=T+1}^{S} (\phi_t(z_t, w_t) - \phi_t(\hat{x}_t, \hat{u}_t)) = \sum_{t=0}^{T} (\phi_t(z_t, w_t) - \phi_t(\hat{x}_t, \hat{u}_t)) + 0 \\
&= F_T((z_t)_{0 \le t \le T}, (w_t)_{0 \le t \le T-1}) - F_T((\hat{x}_t)_{0 \le t \le T}, (\hat{u}_t)_{0 \le t \le T-1})
\end{aligned}
$$

and then taking $S \to +\infty$ we obtain

$$
\begin{aligned}
&\liminf_{S \to +\infty} \sum_{t=0}^{S} (\phi_t(z_t, w_t) - \phi_t(\hat{x}_t, \hat{u}_t)) \\
&= F_T((z_t)_{0 \le t \le T}, (w_t)_{0 \le t \le T-1}) - F_T((\hat{x}_t)_{0 \le t \le T}, (\hat{u}_t)_{0 \le t \le T-1}) > 0
\end{aligned}
$$

which is a contradiction.

For $(\mathscr{P}_e^w)$, using the same reasoning as that of the previous optimality criterion, we obtain

$$
\begin{aligned}
&\limsup_{S \to +\infty} \sum_{t=0}^{S} (\phi_t(z_t, w_t) - \phi_t(\hat{x}_t, \hat{u}_t)) \\
&= F_T((z_t)_{0 \le t \le T}, (w_t)_{0 \le t \le T-1}) - F_T((\hat{x}_t)_{0 \le t \le T}, (\hat{u}_t)_{0 \le t \le T-1}) > 0
\end{aligned}
$$

which is a contradiction.

The reasoning is similar for problems governed by (DI). □

1.4 Necessary Conditions of Optimality

1.4.1 Introduction: A Contribution of Boltyanskii

First we specify the notions of "maximum principle." The **strong maximum principle** for an optimal process $(\underline{\hat{x}}, \underline{\hat{u}})$ is the following property:

$$H_t(\hat{x}_t, \hat{u}_t, p_{t+1}, \lambda_0) = \max\{H_t(\hat{x}_t, u, p_{t+1}, \lambda_0) : u \in U_t\} \tag{1.2}$$

for all $t \in \mathbb{N}$.

Sometimes, the Hamiltonian H_t is replaced by a sum of H_t and a combination of constraint functions with multipliers associated to these constraints.

We call a **weak maximum principle** a condition which is like a first-order necessary condition of optimality for (1.2); for instance

- $D_2 H_t(\hat{x}_t, \hat{u}_t, p_{t+1}, \lambda_0) = 0$ when this partial differential exists and when $\hat{u}_t$ belongs to the interior of U_t.
- $\langle D_2 H_t(\hat{x}_t, \hat{u}_t, p_{t+1}, \lambda_0), u - \hat{u}_t \rangle \leq 0$ when this partial differential exists and when U_t is convex.
- $\partial_2 H_t(\hat{x}_t, \hat{u}_t, p_{t+1}, \lambda_0) \in N_{U_t}(\hat{u}_t)$ when this generalized partial differential exists in the setting of the Clarke calculus (see Appendix B).

As for the strong principles, the Hamiltonian H_t is sometimes replaced by a sum of H_t and a combination of constraint functions with multipliers associated to these constraints.

In his book [30], which is a treatise (probably the first one) on the finite-horizon discrete-time optimal control theory, Boltyanskii explains that a major difference between the continuous-time case and the discrete-time case for optimal control problems is the following one: in continuous time the strong maximum principle is obtained without special assumptions, but in discrete time, it is necessary to add special assumptions to obtain a strong maximum principle. He provides a counterexample in [30] p. 77, and he cites published works which contain false theorems on the strong maximum principle (p. 35 in [30]). Following this contribution, the general directions for the mathematical work on the discrete-time Pontryagin principle are:

- Try to obtain weak maximum principles under assumptions that are as light as possible.
- Understand what are the additional assumptions to make in order to obtain a strong maximum principle.

At this stage, it can be useful to recall that Boltyanskii is the major contributor to the establishment of a Pontryagin principle in the general case of the continuous-time finite-horizon framework, see [76]. Since we speak of history, Michel later provided another proof of the Pontryagin principle in the continuous-time finite-horizon framework in [69] (a paper which is difficult to find without a Mathematical

Reviews number and which is written in French); the proof of Michel is exposed in the book of Alexeev, Tihomirov, and Fomin [1] (an English edition of this book was realized by Plenum, at New York, in 1987).

In his book [30], Boltyanskii established weak maximum principles using a kind of tangent cone (that he calls "copula") and continuously differentiable functions.

1.4.2 The Use of the Multiplier Rule of Halkin

This multiplier rule of Halkin is given in Appendix B; we use it on problems $(\mathcal{F}^e(T, \eta, \hat{x}_T))$ and $(\mathcal{F}^i(T, \eta, \hat{x}_T))$ defined in Sect. 1.3. To do so we translate $(\mathcal{F}^e(T, \eta, \hat{x}_T))$ into a static optimization problem. Note that, in $(\mathcal{F}^e(T, \eta, \hat{x}_T))$, x_0 and x_T are fixed and so they are not unknown variables. We also assume that the sets of admissible controls U_t are defined by equalities and inequalities:

$$U_t = \left(\bigcap_{\alpha=1}^{k^i} \{u \in \mathbb{R}^d : g_t^\alpha(u) \geq 0\} \right) \cap \left(\bigcap_{\beta=1}^{k^e} \{u \in \mathbb{R}^d : h_t^\beta(u) = 0\} \right) \tag{1.3}$$

for all $t \in \mathbb{N}$. We arbitrarily fix $T \in \mathbb{N}$, $T \geq 2$ and we set

$$\tilde{g}^0(x_1, \ldots, x_{T-1}, u_0, \ldots, u_{T-1}) := \phi_0(\eta, u_0) + \sum_{t=1}^{T-1} \phi_t(x_t, u_t), \tag{1.4}$$

and for all $\alpha \in \{1, \ldots, k^i\}$, and for all $t \in \{1, \ldots, T-1\}$, we set

$$\tilde{g}_t^\alpha(x_1, \ldots, x_{T-1}, u_0, \ldots, u_{T-1}) := g_t^\alpha(u_t), \tag{1.5}$$

and for all $\beta \in \{1, \ldots, k^e\}$, and for all $t \in \{1, \ldots, T-1\}$, we set

$$\tilde{h}_t^\beta(x_1, \ldots, x_{T-1}, u_0, \ldots, u_{T-1}) := h_t^\beta(u_t), \tag{1.6}$$

and for all $j \in \{1, \ldots, n\}$,

$$\psi_1^j(x_1, \ldots, x_{T-1}, u_0, \ldots, u_{T-1}) := f_0^j(\eta, u_0) - x_1^j \tag{1.7}$$

and $j \in \{1, \ldots, n\}$ and for all $t \in \{1, \ldots, T-1\}$,

$$\psi_t^j(x_1, \ldots, x_{T-1}, u_0, \ldots, u_{T-1}) := f_{t-1}^j(x_{t-1}, u_{t-1}) - x_t^j \tag{1.8}$$

and also $j \in \{1, \ldots, n\}$,

$$\psi_T^j(x_1, \ldots, x_{T-1}, u_0, \ldots, u_{T-1}) := f_{T-1}^j(x_{T-1}, u_{T-1}) - \hat{x}_T^j \tag{1.9}$$

And so problem $(\mathscr{F}^e(T,\eta,\hat{x}_T))$ can be translated into the following form:

$$(\mathscr{F}_1^e(T,\eta,\hat{x}_T)) \begin{cases} \text{Maximize } \tilde{g}^0(x_1,\dots,x_{T-1},u_0,\dots,u_{T-1}) \\ \quad \text{when} \quad \forall t \in \{1,\dots,T-1\}, x_t \in X_t \\ \quad \forall j \in \{1,\dots,n\}, \forall t \in \{1,\dots,T\}, \\ \qquad \psi_t^j(x_1,\dots,x_{T-1},u_0,\dots,u_{T-1}) = 0. \\ \quad \forall \alpha \in \{1,\dots,k^i\}, \forall t \in \{1,\dots,T-1\}, \\ \qquad \tilde{g}_t^\alpha(x_1,\dots,x_{T-1},u_0,\dots,u_{T-1}) \geq 0 \\ \quad \forall \beta \in \{1,\dots,k^e\}, \forall t \in \{1,\dots,T-1\}, \\ \qquad \tilde{h}_t^\beta(x_1,\dots,x_{T-1},u_0,\dots,u_{T-1}) = 0 \end{cases}$$

Since the sets X_t are open, the conditions $x_t \in X_t$ are not constraints.

We see that this problem is in the form of problem $(\mathscr{M})$ in Appendix B. The multiplier associated to the criterion is denoted by λ_0^T, the multiplier associated to the constraint $\tilde{g}_t^\alpha \geq 0$ is denoted by $\lambda_{\alpha,t}^T$, the multiplier associated to the constraint $\tilde{h}_{t,T}^\beta = 0$ is denoted by $\mu_{\beta,t}^T$, and the multiplier associated to the constraint $\psi_t^j = 0$ is denoted by $p_{t+1,j}^T$. We set $p_{t+1}^T := \sum_{j=1}^n p_{t+1,j}^T e_j^* \in \mathbb{R}^{n*}$ where $(e_j^*)_{1\leq j\leq n}$ is the dual basis of the canonical basis of $\mathbb{R}^n$. And then the generalized Lagrangian of this problem is

$$\mathscr{G}(x_1,\dots,x_{T-1},u_0,\dots,u_{T-1},\lambda_0^T,\lambda_{1,1}^T,\dots,\lambda_{k^i,T-1}^T,\mu_{1,1}^T,\dots,\mu_{k^e,T-1}^T,p_{1,1}^T,\dots,p_{T,n}^T)$$
$$= \lambda_0^T\left(\phi_0(\eta,u_0) + \sum_{t=1}^{T-1}\phi_t(x_t,u_t)\right) + \sum_{\alpha=1}^{k^i}\sum_{t=1}^{T-1}\lambda_{\alpha,t}^T g_t^\alpha(u_t) + \sum_{\beta=1}^{k^e}\sum_{t=1}^{T-1}\mu_{\beta,t}^T h_t^\beta(u_t)$$
$$+ \sum_{t=0}^{T-1}\langle p_{t+1}^T, f_t(x_t,u_t) - x_{t+1}\rangle.$$

As preliminary calculations, if all these functions are differentiable, the differential of the generalized Lagrangian with respect to x_t is

$$D_{x_t}\mathscr{G}(x_1,\dots,x_{T-1},u_0,\dots,u_{T-1},\lambda_0^T,\lambda_{1,1}^T,\dots,\lambda_{k^i,T-1}^T,\mu_{1,1}^T,\dots,\mu_{k^e,T-1}^T,p_{1,1}^T,\dots,p_{T,T}^n)$$
$$= \lambda_0^T D_1\phi_t(x_t,u_t) + p_{t+1}^T \circ D_1 f_t(x_t,u_t) - p_t,$$

and

$$D_{u_t}\mathscr{G}(x_1,\dots,x_{T-1},u_0,\dots,u_{T-1},\lambda_0^T,\lambda_{1,1}^T,\dots,\lambda_{k^i,T-1}^T,\mu_{1,1}^T,\dots,\mu_{k^e,T-1}^T,p_{1,1}^T,\dots,p_{T,T}^n)$$
$$= \lambda_0^T D_2\phi_t(x_t,u_t) + p_{t+1}^T \circ D_2 f_t(x_t,u_t) + \sum_{\alpha=1}^{k^i}\lambda_{\alpha,t}^T Dg_t^\alpha(u_t) + \sum_{\beta=1}^{k^e}\mu_{\beta,t}^T Dh_t^\beta(u_t).$$

And we can state the following result.

Proposition 1.3. *Let $(\underline{\hat{x}}, \underline{\hat{u}})$ be a solution of $(\mathscr{P}_e^n)$, or of $(\mathscr{P}_e^s)$, or of $(\mathscr{P}_e^o)$, or of $(\mathscr{P}_e^w)$. We assume that U_t is defined by (1.3). We also assume that, for all $t \in \mathbb{N}$, for all $\alpha \in \{1, \dots, k^i\}$, and for all $\beta \in \{1, \dots, k^e\}$, the functions ϕ_t, f_t, g_t^α, h_t^β are continuous on a neighborhood of $(\hat{x}_t, \hat{u}_t)$ and they are Fréchet differentiable at $(\hat{x}_t, \hat{u}_t)$. Then, for all $T \in \mathbb{N}$, $T \geq 2$, there exist real numbers λ_0^T, $\lambda_{\alpha,t}^T$ (for all $\alpha \in \{1, \dots, k^i\}$ and for all $t \in \{0, \dots, T-1\}$), $\mu_{\beta,t}^T$ (for all $\beta \in \{1, \dots, k^e\}$ and for all $t \in \{0, \dots, T-1\}$), and $p_{t,j}^T$ (for all $j \in \{1, \dots, n\}$ and for all $t \in \{1, \dots, T\}$) which satisfy the following relations:*

(i) λ_0^T, $(\lambda_{\alpha,t}^T)_{0 \leq t \leq T-1, 1 \leq \alpha \leq k^i}$, $(\mu_{\beta,t}^T)_{0 \leq t \leq T-1, 1 \leq \beta \leq k^e}$, $(p_t^T)_{1 \leq t \leq T}$ *are not simultaneously equal to zero.*
(ii) $\lambda_0^T \geq 0$, $\lambda_{\alpha,t}^T \geq 0$ *for all* $\alpha \in \{1, \dots, k^i\}$ *and for all* $t \in \{0, \dots, T-1\}$.
(iii) $\lambda_{\alpha,t}^T g_t^\alpha(\hat{u}_t) = 0$ *for all* $\alpha \in \{1, \dots, k^i\}$ *and for all* $t \in \{0, \dots, T-1\}$.
(iv) $p_t = D_1 H_t(\hat{x}_t, \hat{u}_t, p_{t+1}^T, \lambda_0^T) = 0$ *for all* $t \in \{1, \dots, T-1\}$.
(v) $D_2 H_t(\hat{x}_t, \hat{u}_t, p_{t+1}^T, \lambda_0^T) + \sum_{\alpha=1}^{k^i} \lambda_{\alpha,t}^T D g_t^\alpha(\hat{u}_t) + \sum_{\beta=1}^{k^e} \mu_{\beta,t}^T D h_t^\beta(\hat{u}_t) = 0$ *for all* $t \in \{0, \dots, T-1\}$.

Moreover when, for all $t \in \mathbb{N}$, $Dg_t^1(\hat{u}_t), \dots, Dg_t^{k^i}(\hat{u}_t), Dh_t^1(\hat{u}_t), \dots, Dh_t^{k^e}(\hat{u}_t)$ are linearly independent we can assert that λ_0^T and $(p_t^T)_{1 \leq t \leq T}$ are not simultaneously equal to zero.

Proof. Using Proposition 1.2 the restriction $((\hat{x}_0, \dots, \hat{x}_T), (\hat{u}_0, \dots, \hat{u}_{T-1}))$ is a solution of problem $(\mathscr{F}^e(T, \eta, \hat{x}_T))$, and then $((\hat{x}_1, \dots, \hat{x}_{T-1}), (\hat{u}_0, \dots, \hat{u}_{T-1}))$ is a solution of problem $(\mathscr{F}_1^e(T, \eta, \hat{x}_T))$. Then using Theorem B.1 in Appendix, we obtain the result. □

In the special case where $\hat{u}_t$ belongs to the interior of U_t we obtain the following more simple statement.

Corollary 1.1. *Let $(\underline{\hat{x}}, \underline{\hat{u}})$ be a solution of $(\mathscr{P}_e^n)$, or of $(\mathscr{P}_e^s)$, or of $(\mathscr{P}_e^o)$, or of $(\mathscr{P}_e^w)$. We assume that $\hat{u}_t \in \operatorname{int} U_t$ for all $t \in \mathbb{N}$. We also assume that, for all $t \in \mathbb{N}$, for all $\alpha \in \{1, \dots, k^i\}$, and for all $\beta \in \{1, \dots, k^e\}$, the functions ϕ_t, f_t, g_t^α, h_t^β are continuous on a neighborhood of $(\hat{x}_t, \hat{u}_t)$ and they are Fréchet differentiable at $(\hat{x}_t, \hat{u}_t)$.*

Then, for all $T \in \mathbb{N}$, $T \geq 2$, there exist real numbers λ_0^T, $p_{t,j}^T$ (for all $j \in \{1, \dots, n\}$ and for all $t \in \{1, \dots, T\}$) which satisfy the following relations where $p_{t+1}^T := \sum_{j=1}^n p_{t+1,j}^T e_j^* \in \mathbb{R}^{n*}$.

(i) λ_0^T, $(p_t^T)_{1 \leq t \leq T}$ *are not simultaneously equal to zero.*
(ii) $\lambda_0^T \geq 0$.
(iii) $p_t = D_1 H_t(\hat{x}_t, \hat{u}_t, p_{t+1}^T, \lambda_0^T)$ *for all* $t \in \{1, \dots, T-1\}$.
(iv) $D_2 H_t(\hat{x}_t, \hat{u}_t, p_{t+1}^T, \lambda_0^T) = 0$ *for all* $t \in \{0, \dots, T-1\}$.

Now we consider problem $(\mathscr{F}^i(T, \eta, \hat{x}_T))$. Using the functions defined by (1.4)–(1.9) we can translate $(\mathscr{F}^i(T, \eta, \hat{x}_T))$ into the following problem:

$$(\mathscr{F}_1^i(T,\eta,\hat{x}_T)) \begin{cases} \text{Maximize } \tilde{g}^0(x_1,\dots,x_{T-1},u_0,\dots,u_{T-1}) \\ \quad \text{when} \quad \forall t \in \{1,\dots,T-1\}, x_t \in X_t \\ \qquad \forall j \in \{1,\dots,n\}, \forall t \in \{1,\dots,T\}, \\ \qquad\qquad \psi_t^j(x_1,\dots,x_{T-1},u_0,\dots,u_{T-1}) \geq 0 \\ \qquad \forall \alpha \in \{1,\dots,k^i\}, \forall t \in \{1,\dots,T-1\}, \\ \qquad\qquad \tilde{g}_t^\alpha(x_1,\dots,x_{T-1},u_0,\dots,u_{T-1}) \geq 0 \\ \qquad \forall \beta \in \{1,\dots,k^e\}, \forall t \in \{1,\dots,T-1\}, \\ \qquad\qquad \tilde{h}_t^\beta(x_1,\dots,x_{T-1},u_0,\dots,u_{T-1}) = 0. \end{cases}$$

The sets X_t being open, the conditions $x_t \in X_t$ are not constraints.

And we can state the following result.

Proposition 1.4. *Let $(\underline{\hat{x}}, \underline{\hat{u}})$ be a solution of $(\mathscr{P}_i^n)$, or of $(\mathscr{P}_i^s)$, or of $(\mathscr{P}_i^o)$, or of $(\mathscr{P}_i^w)$. We assume that U_t is defined by (1.3) for all $t \in \mathbb{N}$. We also assume that, for all $t \in \mathbb{N}$, for all $\alpha \in \{1,\dots,k^i\}$, and for all $\beta \in \{1,\dots,k^e\}$, the functions ϕ_t, f_t, g_t^α, h_t^β are continuous on a neighborhood of $(\hat{x}_t, \hat{u}_t)$ and they are Fréchet differentiable at $(\hat{x}_t, \hat{u}_t)$. Then, for all $T \in \mathbb{N}$, $T \geq 2$, there exist real numbers λ_0^T, $p_{t,j}^T$ (for all $j \in \{1,\dots,n\}$ and for all $t \in \{1,\dots,T\}$), $\lambda_{\alpha,t}^T$ (for all $\alpha \in \{1,\dots,k^i\}$ and for all $t \in \{0,\dots,T-1\}$), and $\mu_{\beta,t}^T$ (for all $\beta \in \{1,\dots,k^e\}$ and for all $t \in \{0,\dots,T-1\}$) which satisfy the following relations:*

(i) λ_0^T, $(p_t^T)_{1\leq t\leq T}$, $(\lambda_{\alpha,t}^T)_{0\leq t\leq T-1, 1\leq\alpha\leq k^i}$, $(\mu_{\beta,t}^T)_{0\leq t\leq T-1,1\leq\beta\leq k^e}$ *are not simultaneously equal to zero.*

(ii) $\lambda_0^T \geq 0$, $p_t^T \geq 0$ *for all* $t \in \{1,\dots,T\}$, *and* $\lambda_{\alpha,t}^T \geq 0$ *for all* $\alpha \in \{1,\dots,k^i\}$ *and for all* $t \in \{0,\dots,T-1\}$.

(iii) $\langle p_{t+1}^T, f_t(\hat{x}_t,\hat{u}_t) - \hat{x}_{t+1}\rangle = 0$ *and* $\lambda_{\alpha,t}^T g_t^\alpha(\hat{u}_t) = 0$ *for all* $\alpha \in \{1,\dots,k^i\}$ *and for all* $t \in \{0,\dots,T-1\}$.

(iv) $p_t = D_1 H_t(\hat{x}_t, \hat{u}_t, p_{t+1}^T, \lambda_0^T) = 0$ *for all* $t \in \{1,\dots,T-1\}$.

(v) $D_2 H_t(\hat{x}_t,\hat{u}_t,p_{t+1}^T,\lambda_0^T) + \sum_{\alpha=1}^{k^i} \lambda_{\alpha,t}^T Dg_t^\alpha(\hat{u}_t) + \sum_{\beta=1}^{k^e} \mu_{\beta,t}^T Dh_t^\beta(\hat{u}_t) = 0$, *for all* $t \in \{0,\dots,T-1\}$.

Moreover, when for all $t \in \mathbb{N}$, $Dg_t^1(\hat{u}_t), \dots, Dg_t^{k^i}(\hat{u}_t)$, $Dh_t^1(\hat{u}_t), \dots, Dh_t^{k^e}(\hat{u}_t)$ are linearly independent we can assert that λ_0^T and $(p_t^T)_{1\leq t\leq T}$ are not simultaneously equal to zero.

The proof is similar to that of Proposition 1.2 replacing $(\mathscr{F}^e(T,\eta,\hat{x}_T))$ by $(\mathscr{F}^i(T,\eta,\hat{x}_T))$ and $(\mathscr{F}_1^e(T,\eta,\hat{x}_T))$ by $(\mathscr{F}_1^i(T,\eta,\hat{x}_T))$.

In the special case where $\hat{u}_t$ belongs to the interior of U_t we easily obtain the following corollary.

Corollary 1.2. *Let $(\underline{\hat{x}}, \underline{\hat{u}})$ be a solution of $(\mathscr{P}_i^n)$, or of $(\mathscr{P}_i^s)$, or of $(\mathscr{P}_i^o)$, or of $(\mathscr{P}_i^w)$. We assume that $\hat{u}_t \in \text{int} U_t$ for all $t \in \mathbb{N}$. We also assume that, for all $t \in \mathbb{N}$, for all $\alpha \in \{1,\dots,k^i\}$, and for all $\beta \in \{1,\dots,k^e\}$, the functions ϕ_t, f_t, g_t^α, h_t^β are continuous on a neighborhood of $(\hat{x}_t,\hat{u}_t)$ and they are Fréchet differentiable at $(\hat{x}_t,\hat{u}_t)$.*

Then, for all $T \in \mathbb{N}$, $T \geq 2$, there exist real numbers λ_0^T and $p_{t,j}^T$ (for all $j \in \{1, \ldots, n\}$ and for all $t \in \{1, \ldots, T\}$) which satisfy the following relations:

(i) λ_0^T, $(p_t^T)_{1 \leq t \leq T}$ *are not simultaneously equal to zero.*
(ii) $\lambda_0^T \geq 0$, *and* $p_t^T \geq 0$ *for all* $t \in \{1, \ldots, T\}$.
(iii) $\langle p_{t+1}^T, f_t(\hat{x}_t, \hat{u}_t) - \hat{x}_{t+1} \rangle = 0$ *for all* $t \in \{0, \ldots, T-1\}$.
(iv) $p_t = D_1 H_t(\hat{x}_t, \hat{u}_t, p_{t+1}^T, \lambda_0^T) = 0$ *for all* $t \in \{1, \ldots, T-1\}$.
(v) $D_2 H_t(\hat{x}_t, \hat{u}_t, p_{t+1}^T, \lambda_0^T) = 0$ *for all* $t \in \{0, \ldots, T-1\}$.

1.4.3 The Use of the Multiplier Rule of Clarke

In this subsection, instead of using the multiplier rule of Halkin we use that of Clarke. In the following result we use notions from Clarke's calculus. These notions are defined in Appendix B.

Proposition 1.5. *Let $(\underline{\hat{x}}, \underline{\hat{u}})$ be a solution of $(\mathscr{P}_e^n)$, or of $(\mathscr{P}_e^s)$, or of $(\mathscr{P}_e^o)$, or of $(\mathscr{P}_e^w)$. We assume that the following conditions are fulfilled:*

(a) For all $t \in \mathbb{N}$, ϕ_t is Lipschitzian on a neighborhood of $(\hat{x}_t, \hat{u}_t)$ and regular at $(\hat{x}_t, \hat{u}_t)$.
(b) For all $t \in \mathbb{N}$, f_t is strictly differentiable at $(\hat{x}_t, \hat{u}_t)$.
(c) For all $t \in \mathbb{N}$, U_t is closed and Clarke-regular at $\hat{u}_t$.

Then, for all $T \in \mathbb{N}$, $T \geq 2$, there exist $\lambda_0^T \in \mathbb{R}$, $p_1^T, \ldots, p_T^T \in \mathbb{R}^{n*}$ which satisfy the following conditions:

(i) $\lambda_0^T, p_1^T, \ldots, p_T^T$ *are not simultaneously equal to zero.*
(ii) $\lambda_0^T \geq 0$.
(iii) For all $t \in \{1, \ldots, T-1\}$, there exists $\varphi_t^T \in \partial_1 \phi_t(\hat{x}_t, \hat{u}_t)$ such that
$p_t^T = p_{t+1}^T \circ D_1 f_t(\hat{x}_t, \hat{u}_t) + \lambda_0^T \varphi_t^T$
(or $p_t^T \in p_{t+1}^T \circ D_1 f_t(\hat{x}_t, \hat{u}_t) + \lambda_0^T \partial_1 \phi_t(\hat{x}_t, \hat{u}_t)$).
(iv) For all $t \in \{0, \ldots, T-1\}$, there exists $\gamma_t^T \in \partial_2 \phi_t(\hat{x}_t, \hat{u}_t)$ such that
$\langle \lambda_0^T \gamma_t^T + p_{t+1}^T \circ D_2 f_t(\hat{x}_t, \hat{u}_t), v_t \rangle \leq 0$ *for all $v_t \in T_{U_t}(\hat{u}_t)$ the tangent cone of Clarke of U_t at $\hat{u}_t$.*

Proof. Using (1.4), (1.7), and (1.9) and denoting $\psi_t := (\psi_t^1, \ldots, \psi_t^n)$, problem $(\mathscr{F}^e(T, \eta, \hat{x}_T))$ can be rewritten as follows:

$$(\mathscr{F}_2^e(T,\eta,\hat{x}_T)) \begin{cases} \text{Maximize } \tilde{g}^0(x_1, \ldots, x_{T-1}, u_0, \ldots, u_{T-1}) \\ \quad \text{when } \quad \forall t \in \{0, \ldots, T-1\}, \psi_t(x_1, \ldots, x_{T-1}, u_0, \ldots, u_{T-1}) = 0 \\ \qquad\qquad\quad (x_1, \ldots, x_{T-1}, u_0, \ldots, u_{T-1}) \in S \end{cases}$$

where $S := X_1 \times \ldots \times X_{T-1} \times U_0 \times \ldots \times U_{T-1}$.

Proceeding as in Proposition 1.2, for all $T \in \mathbb{N}$, $T \geq 2$, the restriction $(\hat{x}_1, \ldots, \hat{x}_{T-1}, \hat{u}_0, \ldots, \hat{u}_{T-1})$ is a solution of $(\mathscr{F}_2^e(T, \eta, \hat{x}_T))$. Under our assumptions, the assumptions of Theorem B.9 in Appendix B are fulfilled and so we obtain the existence of λ_0^T and p_{t+1}^T for all $t \in \{0, \dots, T-1\}$ such that

(i) $\lambda_0^T, p_1^T, \dots, p_T^T$ are not simultaneously equal to zero.
(ii) $\lambda_0^T \geq 0$.
(iii') $0 \in \partial_I L^C(\hat{x}_1, \dots, \hat{x}_{T-1}, \hat{u}_0, \dots, \hat{u}_{T-1}, p_1^T, \dots, p_T^T, \lambda_0^T, k)$ where ∂_I denotes the Clarke differential with respect to $(x_1, \dots, x_{T-1}, u_0, \dots, u_{T-1})$.

Now it remains to show why (iii') implies the conclusions (iii) and (iv) of the statement. The condition (iii') implies (cf. the comments after Theorem B.9 in Appendix B)

$$0 \in \partial_I \left(\sum_{t=0}^{T-1} (H_t(\hat{x}_t, \hat{u}_t, p_{t+1}^T, \lambda_0^T) - \langle p_{t+1}^T, \hat{x}_{t+1} \rangle) \right) - \hat{N}_P \tag{1.10}$$

where $\hat{N}_P := N_P(\hat{x}_1, \dots, \hat{x}_{T-1}, \hat{u}_0, \dots, \hat{u}_{T-1})$ is the normal cone of P at $(\hat{x}_1, \dots, \hat{x}_{T-1}, \hat{u}_0, \dots, \hat{u}_{T-1})$ (see Appendix B). We note that

$$\lambda_0^T \phi - t(x_t, u_t) = H_t(x_t, u_t, p_{t+1}^T, \lambda_0^T) - \langle p_{t+1}^T, x_{t+1} \rangle.$$

Using Theorem B.7 we have $\hat{N}_P = \prod_{t=1}^{T-1} N_{X_t}(\hat{x}_t) \times \prod_{t=0}^{T-1} N_{U_t}(\hat{u}_t)$. The set X_t being open, its tangent cone at $\hat{x}_t$ is $\mathbb{R}^n$ and consequently $N_{X_t}(\hat{x}_t) = \{0\}$ and so we have

$$\hat{N}_P = \prod_{t=1}^{T-1} \{0\} \times \prod_{t=0}^{T-1} N_{U_t}(\hat{u}_t). \tag{1.11}$$

Under the assumptions of regularity and of strict differentiability we obtain (cf. Theorem B.5 in Appendix B)

$$\left.\begin{aligned} &\partial_I \left(\sum_{t=0}^{T-1} (H_t(\hat{x}_t, \hat{u}_t, p_{t+1}^T, \lambda_0^T) - \langle p_{t+1}^T, \hat{x}_{t+1} \rangle) \right) - \hat{N}_P \\ &\subset \prod_{t=1}^{T-1} \left(\partial_{x_t} \left(\sum_{s=0}^{T-1} (H_s(\hat{x}_s, \hat{u}_s, p_{s+1}^T, \lambda_0^T) - \langle p_{s+1}^T, \hat{x}_{s+1} \rangle) \right) - \{0\} \right. \\ &\times \prod_{t=1}^{T-1} \left(\partial_{u_t} \left(\sum_{s=0}^{T-1} (H_s(\hat{x}_s, \hat{u}_s, p_{s+1}^T, \lambda_0^T) - \langle p_{s+1}^T, \hat{x}_{s+1} \rangle) \right) - N_{U_t}(\hat{u}_t) \right. . \end{aligned}\right\} \tag{1.12}$$

Using Theorem B.6 in Appendix B, we have, for all $t \in \{1, \dots, T-1\}$,

$$\left.\begin{aligned} &\partial_{x_t} \left(\sum_{s=0}^{T-1} (H_s(\hat{x}_s, \hat{u}_s, p_{s+1}^T, \lambda_0^T) - \langle p_{s+1}^T, \hat{x}_{s+1} \rangle) \right) \\ &= \partial_1 H_t(\hat{x}_t, \hat{u}_t, p_{t+1}^T, \lambda_0^T) - p_t^T, \end{aligned}\right\} \tag{1.13}$$

and

$$\left.\begin{aligned}&\partial_{u_t}\left(\sum_{s=0}^{T-1}(H_s(\hat{x}_s,\hat{u}_s,p_{s+1}^T,\lambda_0^T)-\langle p_{s+1}^T,\hat{x}_{s+1}\rangle)\right)-N_{U_t}(\hat{u}_t)\\&=\partial_2 H_t(\hat{x}_t,\hat{u}_t,p_{t+1}^T,\lambda_0^T)-N_{U_t}(\hat{u}_t).\end{aligned}\right\}\tag{1.14}$$

From (1.10), (1.12), and (1.13), we obtain

$$0\in\partial_1 H_t(\hat{x}_t,\hat{u}_t,p_{t+1}^T,\lambda_0^T)-p_t^T$$

which implies $p_t^T\in\partial_1 H_t(\hat{x}_t,\hat{u}_t,p_{t+1}^T,\lambda_0^T)$ that gives (iii).

From (1.10), (1.12), and (1.14) we obtain

$$0\in\partial_2 H_t(\hat{x}_t,\hat{u}_t,p_{t+1}^T,\lambda_0^T)-N_{U_t}(\hat{u}_t)$$

which implies $\partial_2 H_t(\hat{x}_t,\hat{u}_t,p_{t+1}^T,\lambda_0^T)\cap N_{U_t}(\hat{u}_t)\neq\emptyset$. Since

$$\partial_2 H_t(\hat{x}_t,\hat{u}_t,p_{t+1}^T,\lambda_0^T)=\lambda_0^T\partial_2\phi_t(\hat{x}_t,\hat{u}_t)+p_{t+1}^T\circ D_2 f_t(\hat{x}_t,\hat{u}_t),$$

there exists $\gamma_t^T\in\partial_2\phi_t(\hat{x}_t,\hat{u}_t)$ such that

$$\lambda_0^T\gamma_t^T+p_{t+1}^T\circ D_2 f_t(\hat{x}_t,\hat{u}_t)\in N_{U_t}(\hat{u}_t)$$

and since the tangent cone is the dual cone to the normal cone, we obtain that, for all $v\in T_{U_t}(\hat{u}_t)$, we have

$$\langle\lambda_0^T\gamma_t^T+p_{t+1}^T\circ D_2 f_t(\hat{x}_t,\hat{u}_t),v\rangle\leq 0$$

that is (iv). And so the proposition is proven. □

When $\hat{u}_t\in\text{int}U_t$ we obtain the following corollary.

Corollary 1.3. *Let $(\underline{\hat{x}},\underline{\hat{u}})$ be a solution of $(\mathcal{P}_e^n)$, or of $(\mathcal{P}_e^s)$, or of $(\mathcal{P}_e^o)$, or of $(\mathcal{P}_e^w)$. We assume that the conditions (a) and (b) of Proposition 1.5 are fulfilled and moreover that $\hat{u}_t\in\text{int}U_t$ for all $t\in\mathbb{N}$. Then, for all $T\in\mathbb{N}$, $T\geq 2$, there exist $\lambda_0^T\in\mathbb{R}$, $p_1^T,\ldots,p_T^T\in\mathbb{R}^{n*}$ which satisfy the following conditions:*

(i) λ_0^T, $p_1^T,\ldots,p_T^T$ are not simultaneously equal to zero.
(ii) $\lambda_0^T\geq 0$.
(iii) For all $t\in\{1,\ldots,T-1\}$, there exists $\varphi_t^T\in\partial_1\phi_t(\hat{x}_t,\hat{u}_t)$ such that
$p_t^T=p_{t+1}^T\circ D_1 f_t(\hat{x}_t,\hat{u}_t)+\lambda_0^T\varphi_t^T$
(or $p_t^T\in p_{t+1}^T\circ D_1 f_t(\hat{x}_t,\hat{u}_t)+\lambda_0^T\partial_1\phi_t(\hat{x}_t,\hat{u}_t)$).
(iv) For all $t\in\{0,\ldots,T-1\}$, $0\in\lambda_0^T\partial_2\phi_t(\hat{x}_t,\hat{u}_t)+p_{t+1}^T\circ D_2 f_t(\hat{x}_t,\hat{u}_t)$.

Proof. The only difference with Proposition 1.5 is the conclusion (iv). When $\hat{u}_t\in\text{int}U_t$ we have $T_{U_t}(\hat{u}_t)=\mathbb{R}^d$, and then the conclusion (iv) of Proposition 1.5

becomes $\langle \lambda_0^T \gamma_t^T + p_{t+1}^T \circ D_2 f_t(\hat{x}_t, \hat{u}_t), v \rangle \leq 0$ for all $v \in \mathbb{R}^d$ which implies $0 \in \lambda_0^T \gamma_t^T + p_{t+1}^T \circ D_2 f_t(\hat{x}_t, \hat{u}_t)$. □

Now we consider the systems which are governed by (DI) instead of (DE). In this case the restriction $(\hat{x}_1, \ldots, \hat{x}_{T-1}, \hat{u}_0, \ldots, \hat{u}_{T-1})$ of a solution $(\underline{\hat{x}}, \underline{\hat{u}})$ of $(\mathcal{P}_i^n)$, or of $(\mathcal{P}_i^s)$, or of $(\mathcal{P}_i^o)$, or of $(\mathcal{P}_i^w)$ is a solution of the following problem:

$$(\mathcal{F}_2^i(T, \eta, \hat{x}_T)) \begin{cases} \text{Maximize } \tilde{g}^0(x_1, \ldots, x_{T-1}, u_0, \ldots, u_{T-1}) \\ \text{when } \quad \forall t \in \{0, \ldots, T-1\}, \psi_t(x_1, \ldots, x_{T-1}, u_0, \ldots, u_{T-1}) \geq 0 \\ \qquad\qquad (x_1, \ldots, x_{T-1}, u_0, \ldots, u_{T-1}) \in S. \end{cases}$$

Reasoning like in the proof of Proposition 1.5 we obtain the following result.

Proposition 1.6. *Let $(\underline{\hat{x}}, \underline{\hat{u}})$ be a solution of $(\mathcal{P}_i^n)$, or of $(\mathcal{P}_i^s)$, or of $(\mathcal{P}_i^o)$, or of $(\mathcal{P}_i^w)$. We assume that the following conditions are fulfilled:*

(a) For all $t \in \mathbb{N}$, ϕ_t is Lipschitzian on a neighborhood of $(\hat{x}_t, \hat{u}_t)$ and regular at $(\hat{x}_t, \hat{u}_t)$.
(b) For all $t \in \mathbb{N}$, f_t is strictly differentiable at $(\hat{x}_t, \hat{u}_t)$.
(c) For all $t \in \mathbb{N}$, U_t is closed and Clarke-regular at $\hat{u}_t$.

Then, for all $T \in \mathbb{N}$, $T \geq 2$, there exist $\lambda_0^T \in \mathbb{R}$, $p_1^T, \ldots, p_T^T \in \mathbb{R}^{n*}$ which satisfy the following conditions:

(i) $\lambda_0^T, p_1^T, \ldots, p_T^T$ are not simultaneously equal to zero.
(ii) $\lambda_0^T \geq 0$, for all $t \in \{0, \ldots, T-1\}$, $p_{t+1}^T \geq 0$ and $\langle p_{t+1}^T, f_t(\hat{x}_t, \hat{u}_t) - \hat{x}_{t+1} \rangle = 0$.
(iii) For all $t \in \{1, \ldots, T-1\}$, there exists $\varphi_t^T \in \partial_1 \phi_t(\hat{x}_t, \hat{u}_t)$ such that $p_t^T = p_{t+1}^T \circ D_1 f_t(\hat{x}_t, \hat{u}_t) + \lambda_0^T \varphi_t^T$ (or $p_t^T \in p_{t+1}^T \circ D_1 f_t(\hat{x}_t, \hat{u}_t) + \lambda_0^T \partial_1 \phi_t(\hat{x}_t, \hat{u}_t)$).
(iv) For all $t \in \{0, \ldots, T-1\}$, there exists $\gamma_t^T \in \partial_2 \phi_t(\hat{x}_t, \hat{u}_t)$ such that $\langle \lambda_0^T \gamma_t^T + p_{t+1}^T \circ D_2 f_t(\hat{x}_t, \hat{u}_t), v_t \rangle \leq 0$ for all $v_t \in T_{U_t}(\hat{u}_t)$ the tangent cone of Clarke of U_t at $\hat{u}_t$.

And when $\hat{u}_t \in \text{int}U_t$ we obtain the following corollary.

Corollary 1.4. *Let $(\underline{\hat{x}}, \underline{\hat{u}})$ be a solution of $(\mathcal{P}_i^n)$, or of $(\mathcal{P}_i^s)$, or of $(\mathcal{P}_i^o)$, or of $(\mathcal{P}_i^w)$. We assume that the conditions (a) and (b) of Proposition 1.6 are fulfilled and moreover that $\hat{u}_t \in \text{int}U_t$ for all $t \in \mathbb{N}$. Then, for all $T \in \mathbb{N}$, $T \geq 2$, there exist $\lambda_0^T \in \mathbb{R}$, $p_1^T, \ldots, p_T^T \in \mathbb{R}^{n*}$ which satisfy the following conditions:*

(i) $\lambda_0^T, p_1^T, \ldots, p_T^T$ are not simultaneously equal to zero.
(ii) $\lambda_0^T \geq 0$, for all $t \in \{0, \ldots, T-1\}$, $p_{t+1}^T \geq 0$ and $\langle p_{t+1}^T, f_t(\hat{x}_t, \hat{u}_t) - \hat{x}_{t+1} \rangle = 0$.
(iii) For all $t \in \{1, \ldots, T-1\}$, there exists $\varphi_t^T \in \partial_1 \phi_t(\hat{x}_t, \hat{u}_t)$ such that $p_t^T = p_{t+1}^T \circ D_1 f_t(\hat{x}_t, \hat{u}_t) + \lambda_0^T \varphi_t^T$ (or $p_t^T \in p_{t+1}^T \circ D_1 f_t(\hat{x}_t, \hat{u}_t) + \lambda_0^T \partial_1 \phi_t(\hat{x}_t, \hat{u}_t)$).
(iv) For all $t \in \{0, \ldots, T-1\}$, $0 \in \lambda_0^T \partial_2 \phi_t(\hat{x}_t, \hat{u}_t) + p_{t+1}^T \circ D_2 f_t(\hat{x}_t, \hat{u}_t)$.

1.4.4 The Use of a Result of Michel

In [70] Michel considers the following finite-horizon discrete-time optimal control problem for $T \in \mathbb{N}_*$:

$$(\mathscr{P}\mathscr{M}) \begin{cases} \text{Maximize } \sum_{t=0}^{T-1} \phi_t(x_t, u_t) + \varphi_T(x_T) \\ \text{when} \quad x_0 = \eta \\ \qquad \forall t \in \{0, \ldots, T-1\}, \\ \qquad\quad x_{t+1} \in X_{t+1}, u_t \in U_t \\ \qquad\quad x_{t+1} - x_t = \tilde{f}_t(x_t, u_t) \\ \qquad\quad \forall i \in \{1, \ldots, m_t\}, \ \tilde{g}_t^i(x_t, u_t) \geq 0 \\ \qquad\quad \forall i \in \{m_t + 1, \ldots, n_t\}, \ \tilde{g}_t^i(x_t, u_t) = 0 \\ \qquad\quad \forall i \in \{1, \ldots, m_T\}, \ \tilde{g}_T^i(x_T) \geq 0 \\ \qquad\quad \forall i \in \{m_T + 1, \ldots, n_T\}, \ \tilde{g}_T^i(x_T) = 0. \end{cases}$$

In this problem, X_t, U_t, ϕ_t, and η are like ours in Sect. 1.2. The functions φ_T and $\tilde{g}_T^i$ are defined from X_T into $\mathbb{R}$, and the functions $\tilde{g}_t^i$ are defined from $X_t \times U_t$ into $\mathbb{R}$ for all $i \in \{1, \ldots, n_t\}$ and for all $t \in \{0, \ldots, T-1\}$. Michel also introduces the following notation:

$$\begin{aligned} A_t(x_t, x_{t+1}) := \{(\lambda, y_t, z_t) \in \mathbb{R} \times \mathbb{R}^n \times \mathbb{R}^n : \exists u_t \in U_t \ \text{ s.t.} \\ & \lambda \leq \phi_t(x_t, u_t) \\ & y_t = \tilde{f}_t(x_t, u_t) + x_t - x_{t+1} \\ & \forall i \in \{1, \ldots, m_t\}, \ z_t^i \leq \tilde{g}_t^i(x_t, u_t) \\ & \forall i \in \{m_t + 1, \ldots, n_t\}, \ z_t^i = \tilde{g}_t^i(x_t, u_t)\} \end{aligned}$$

and

$$\begin{aligned} B_t(x_t, x_{t+1}) := \{(\lambda, y_t, z_t) \in \mathbb{R} \times \mathbb{R}^n \times \mathbb{R}^n : \exists (u_t, v_t, w_t) \in U_t \times \mathbb{R}^n \times \mathbb{R}^{n_t} \ \text{ s.t.} \\ & \lambda \leq \phi_t(x_t, u_t) \\ & \forall k \in \{1, \ldots, n\}, v_t^k y_t^k = \tilde{f}_t^k(x_t, u_t) + x_t^k - x_{t+1}^k \\ & \forall i \in \{1, \ldots, m_t\}, \ w_t^i z_t^i \leq \tilde{g}_t^i(x_t, u_t) \\ & \forall i \in \{m_t + 1, \ldots, n_t\}, \ w_t^i z_t^i = \tilde{g}_t^i(x_t, u_t)\} \end{aligned}$$

He introduces the following condition that can be called Michel's convexity condition:

For all $t \in \mathbb{N}$, for all $(x_t, x_{t+1}) \in X_t \times X_{t+1}$, $\mathrm{co}A_t(x_t, x_{t+1}) \subset B_t(x_t, x_{t+1})$. Each of the following equivalent conditions implies it:

- For all $t \in \mathbb{N}$, $A_t(x_t, x_{t+1})$ is convex

- Ioffe–Tihomirov' s conditions that we shall use in the next subsection: $\forall t \in \mathbb{N}, \forall x_t \in X_t, \ \ \forall u'_t, u''_t \in U_t, \ \ \forall \theta \in [0,1], \ \ \exists u_t \in U_t$ such that:

$$\begin{array}{l} \phi_t(x_t,u_t) \geq \theta\phi_t(x_t,u'_t) + (1-\theta)\phi_t(x_t,u''_t) \\ f_t(x_t,u_t) = \theta f_t(x_t,u'_t) + (1-\theta) f_t(x_t,u''_t) \\ \tilde{g}^i(x_t,u_t) \geq \theta\tilde{g}^i(x_t,u'_t) + (1-\theta)\tilde{g}^i(x_t,u''_t), \forall i \in \{1,\dots,m_t\} \\ \tilde{g}^i_t(x_t,u_t) = \theta\tilde{g}^i_t(x_t,u'_t) + (1-\theta)\tilde{g}^i_t(x_t,u''_t), \forall i \in \{m_t+1,\dots,n_t\} \end{array}$$

These conditions are themselves implied by the following one: U_t is convex, ϕ_t and $\forall i \in \{1,\dots,m_t\}$, $\tilde{g}^i_t$ concave with respect to u_t, and f_t and $\forall i \in \{m_t+1,\dots,n_t\}$, $\tilde{g}^i_t$ affine with respect to u_t, for all $t \in \mathbb{N}$.

Theorem 1.5. *Let* $(\overline{x}_0,\dots,\overline{x}_T,\overline{u}_0,,\overline{u}_{T-1})$ *be a solution of* $(\mathscr{P}\mathscr{M})$*. We assume that the following conditions are fulfilled:*

(a) For all $t \in \{0,\dots,T-1\}$, X_{t+1} *is an open convex subset of* $\mathbb{R}^n$, U_t *is a nonempty subset of* $\mathbb{R}^d$.
(b) For all $t \in \{0,\dots,T-1\}$, *for all* $i \in \{1,\dots,n_t\}$, *the functions* ϕ_t, $\tilde{f}_t$, $\tilde{g}^i_t$ *are differentiable with respect the first variable, and* φ_T *and* $\tilde{g}^i_T$ *are differentiable.*
(c) For all $t \in \{0,\dots,T-1\}$, *for all* $(x_t,x_{t+1}) \in X_t \times X_{t+1}$, $coA_t(x_t,x_{t+1}) \subset B_t(x_t,x_{t+1})$.

Then there exist real numbers λ_0, b^i_t *(for all* $t \in \{0,\dots,T\}$, *for all* $i \in \{1,\dots,n_t\}$*), and* q^k_{t+1} *(for all* $t \in \{0,\dots,T-1\}$, *for all* $k \in \{1,\dots,n\}$*), not simultaneously equal to zero, which satisfy the following conditions:*

(i) $\lambda_0 \geq 0$,
(ii) $\forall t \in \{0,\dots,T-1\}$, $\forall i \in \{1,\dots,m_t\}$, $b^i_t \geq 0$ *and* $b^i_t\tilde{g}^i_t(\overline{x}_t,\overline{u}_t) = 0$,
(iii) $\forall i \in \{1,\dots,m_T\}$, $b^i_T \geq 0$ *and* $b^i_T\tilde{g}^i_T(\overline{x}_t) = 0$,
(iv) $\forall t \in \{1,\dots,T-1\}$, $\forall k \in \{1,\dots,n\}$, $q^k_t - q^k_{t+1} = \frac{\partial\tilde{H}_t}{\partial x^k}(\overline{x}_t,\overline{u}_t,q_{t+1},\lambda_0,b_t)$,
(v) $\forall k \in \{1,\dots,n\}$, $q^k_T = \lambda_0\frac{\partial\varphi_T}{\partial x^k}(\overline{x}_T) + \sum_{i=1}^{n_T} b^i_T\frac{\partial\tilde{g}^i_T}{\partial x^k}(\overline{x}_T)$,
(vi) $\forall t \in \{0,\dots,T-1\}$, $\tilde{H}_t(\overline{x}_t,\overline{u}_t,q_{t+1},\lambda_0,b_t) = \max_{u\in U_t}\tilde{H}_t(\overline{x}_t,u,q_{t+1},\lambda_0,b_t)$,

where

$$\begin{aligned}\tilde{H}_t(x,u,q,\lambda,b) &:= \lambda\phi_t(x,u) + \sum_{k=1}^{n} q^k\tilde{f}^k_t(x,u) + \sum_{i=1}^{n_t} b^i\tilde{g}^i_t(x,u) \\ &= H_t(x,u,q,\lambda) + \sum_{i=1}^{n_t} b^i\tilde{g}^i_t(x,u).\end{aligned}$$

Now we want to apply this result to our problems $(\mathscr{F}^e(T,\eta,\hat{x}(T)))$. Since the equation of motion of $(\mathscr{P}\mathscr{M})$ is different, we set $\tilde{f}_t(x,u) := f_t(x,u) - x$. Since the constraints $\tilde{g}^i_t(x_t,u_t) \geq 0$ and $\tilde{g}^i_t(x_t,u_t) = 0$ are not present in our problems, we delete them. And so the sets $A_t(x_t,x_{t+1})$ and $B_t(x_t,x_{t+1})$ can be simplified in the following way:

$$A'_t(x_t, x_{t+1}) := \{(\lambda, y_t) \in \mathbb{R} \times \mathbb{R}^n : \exists u_t \in U_t \text{ s.t. } \lambda \leq \phi_t(x_t, u_t),\ y_t = \tilde{f}_t(x_t, u_t) + x_t - x_{t+1}\}$$

and

$$B'_t(x_t, x_{t+1}) := \{(\lambda, y_t) \in \mathbb{R} \times \mathbb{R}^n \times \mathbb{R}^n : \exists (u_t, v_t) \in U_t \times \mathbb{R}^n \times \mathbb{R}^{n_t} \text{ s.t. } \lambda \leq \phi_t(x_t, u_t),\ \forall k \in \{1, \ldots, n\}, v_t^k y_t^k = \tilde{f}_t^k(x_t, u_t) + x_t^k - x_{t+1}^k\}$$

We obtain the following result.

Proposition 1.7. *Let $(\underline{\hat{x}}, \underline{\hat{u}})$ be a solution of $(\mathscr{P}_e^n)$, or of $(\mathscr{P}_e^s)$, or of $(\mathscr{P}_e^o)$, or of $(\mathscr{P}_e^w)$. We assume that the following conditions are fulfilled:*

(a) For all $t \in \mathbb{N}$, X_t is a nonempty open convex subset of $\mathbb{R}^n$ and U_t is a nonempty subset of $\mathbb{R}^d$.
(b) For all $t \in \mathbb{N}$, the functions ϕ_t and f_t are differentiable with respect to the first vector variable.
(c) For all $t \in \mathbb{N}$, for all $(x_t, x_{t+1}) \in X_t \times X_{t+1}$, $coA'_t(x_t, x_{t+1}) \subset B'_t(x_t, x_{t+1})$.

Then for all $T \in \mathbb{N}$, $T \geq 2$, there exist $\lambda_0^T \in \mathbb{R}$ and $p_1^T, \ldots, p_T^T \in \mathbb{R}^{n*}$ which satisfy the following conditions:

(i) λ_0^T, $p_1^T, \ldots, p_T^T$ are not simultaneously equal to zero.
(ii) $\lambda_0^T \geq 0$.
(iii) For all $t \in \{1, \ldots, T-1\}$, $p_t^T = D_1 H_t(\hat{x}_t, \hat{u}_t, p_{t+1}^T, \lambda_0^T)$.
(iv) For all $t \in \{0, \ldots, T-1\}$, $H_t(\hat{x}_t, \hat{u}_t, p_{t+1}^T, \lambda_0^T) = \max_{u \in U_t} H_t(\hat{x}_t, u, p_{t+1}^T, \lambda_0^T)$.

Proof. First it is useful to translate the notation of Theorem 1.5 into our notation. We have $\varphi_T = 0$, $\overline{x}_t = \hat{x}_t$, $\overline{u}_t = \hat{u}_t$. We have just said that $\tilde{f}_t(x, u) := f_t(x, u) - x$, and so the equation of motion $x_{t+1} - x_t = \tilde{f}_t(x_t, u_t)$ is exactly $x_{t+1} = f_t(x_t, u_t)$. About the Hamiltonian, we have the following relation:

$$\tilde{H}_t(x, u, p, \lambda) = H_t(x, u, p, \lambda) - \langle p, x \rangle. \tag{1.15}$$

And so we have

$$D_1 \tilde{H}_t(x, u, p, \lambda) = D_1 H_t(x, u, p, \lambda) - p.$$

Therefore the conclusion (iv) of Theorem 1.5, with $p_t^T = q_t$, becomes

$$p_t^T - p_{t+1}^T = D_1 H_t(\hat{x}_t, \hat{u}_t, p_{t+1}^T, \lambda_0^T) - p_{t+1}^T$$

which implies the conclusion (iii). Again from (1.15), the conclusion (vi) of Theorem 1.5 becomes

$$H_t(\hat{x}_t, \hat{u}_t, p_{t+1}^T, \lambda_0^T) - \langle p_{t+1}^T, \hat{x}_t \rangle = \max_{u \in U_t} (H_t(\hat{x}_t, u, p_{t+1}^T, \lambda_0^T) - \langle p_{t+1}^T, \hat{x}_t \rangle)$$

and since $\langle p_{t+1}^T, \hat{x}_t \rangle$ does not depend of $u \in U_t$ we obtain (iv). □

The finite-horizon Theorem 1.5 is written for systems governed by (DE). To obtain an analogous result for systems which are governed by (DI), it is necessary to go back to the parametrized static optimization theorem of Michel of which Theorem 1.5 is a corollary. The parametrized static optimization problem of Michel is the following:

$$(\mathscr{P}\mathscr{P}) \begin{cases} \text{Maximize } \Gamma^0(x,u) \\ \quad \text{when} \quad (x,u) \in \mathfrak{X} \times \mathfrak{U} \\ \qquad\qquad \forall i \in \{1, \dots, m_0\}, \quad \Gamma^i(x,u) \geq 0 \\ \qquad\qquad \forall i \in \{m_0+1, \dots, m\}, \quad \Gamma^i(x,u) = 0, \end{cases}$$

where $\mathfrak{X} \subset \mathbb{R}^N$, $\mathfrak{U} \subset \mathbb{R}^M$, with $N, M \in \mathbb{N}$. We set

$$\begin{aligned} A(x) := \{(z^0, \dots, z^m) \in \mathbb{R}^{1+m} : \exists u \in \mathfrak{U} \text{ s.t.} \\ \forall i \in \{0, \dots, m_0\}, \quad z^i \leq \Gamma^i(x,u), \\ \forall i \in \{m_0+1, \dots, m\}, \quad z^i = \Gamma^i(x,u)\} \end{aligned}$$

and

$$\begin{aligned} B(x) := \{(z^0, \dots, z^m) \in \mathbb{R}^{1+m} : \exists u \in \mathfrak{U}, \exists v \in \mathbb{R}^{1+m} \text{ s.t.} \\ \forall i \in \{0, \dots, m_0\}, \quad v^i z^i \leq \Gamma^i(x,u), \\ \forall i \in \{m_0+1, \dots, m\}, \quad v^i z^i = \Gamma^i(x,u)\}. \end{aligned}$$

Then we can state the following result of Michel (Theorem 1.4, p. 4, in [70]).

Theorem 1.6. *Let $(\overline{x}, \overline{u})$ be a solution of $(\mathscr{P}\mathscr{P})$. We assume that the following conditions are fulfilled:*

(a) For all $t \in \mathbb{N}$, X_t is nonempty and convex.
(b) For all $t \in \mathbb{N}$, for all $i \in \{0, \dots, m\}$, $\Gamma^i(., \hat{u}_t)$ is continuous on a neighborhood of $\overline{x}$ and it is Fréchet differentiable at $\overline{x}$.
(c) For all $t \in \mathbb{N}$, for all $(x_t, x_{t+1}) \in X_t \times X_{t+1}$, $coA_t(x_t, x_{t+1}) \subset B_t(x_t, x_{t+1})$.

Then there exists $(a_i)_{0 \leq i \leq m} \in \mathbb{R}^{1+m}$ which satisfies the following conditions:

(i) $(a_i)_{0 \leq i \leq m} \in \mathbb{R}^{1+m}$ is not equal to zero.
(ii) For all $i \in \{0, \dots, m_0\}$, $a_i \geq 0$.
(iii) For all $i \in \{1, \dots, m_0\}$, $a_i \Gamma^i(\overline{x}, \overline{u}) = 0$.
(iv) For all $x \in \mathfrak{X}$, $\sum_{i=0}^m a_i \langle D_1 \Gamma^i(\overline{x}, \overline{u}), x - \overline{x} \rangle \leq 0$.
(v) $\sum_{i=0}^m a_i \Gamma^i(\overline{x}, \overline{u}) = \max_{u \in \mathfrak{U}} \sum_{i=0}^m a_i \Gamma^i(\overline{x}, u)$.

Now we want to apply this theorem of Michel to our problems $(\mathcal{F}^i(T,\eta,\hat{x}(T)))$. To do that, we need to introduce the following sets:

$$A_t''(x_t,x_{t+1}) := \{(\lambda,y)\in\mathbb{R}\times\mathbb{R}^n : \exists u_t\in U_t \text{ s.t. } \lambda\le\phi_t(x_t,u_t),\ \ y\le f_t(x_t,u_t)-x_{t+1}\}$$

and

$$B_t''(x_t,x_{t+1}) := \{(\lambda,y)\in\mathbb{R}\times\mathbb{R}^n : \exists u_t\in U_t, \exists v_t\in\mathbb{R}^n \text{ s.t. } \lambda\le\phi_t(x_t,u_t), \forall k\in\{1,\dots,n\}, v^k y^k\le f_t^k(x_t,u_t)-x_{t+1}^k\}$$

Proposition 1.8. *Let $(\underline{\hat{x}},\underline{\hat{u}})$ be a solution of $(\mathcal{P}_i^n)$, or of $(\mathcal{P}_i^s)$, or of $(\mathcal{P}_i^o)$, or of $(\mathcal{P}_i^w)$. We assume that the following conditions are fulfilled:*

(a) For all $t\in\mathbb{N}$, X_t is a nonempty convex subset of $\mathbb{R}^n$, U_t is a nonempty subset of $\mathbb{R}^d$, and $\hat{x}_t\in \operatorname{int}X_t$.
(b) For all $t\in\mathbb{N}$, the functions $\phi_t(.,\hat{u}_t)$ and $f(.,\hat{u}_t)$ are continuous on a neighborhood of $\hat{x}_t$ and Fréchet differentiable at $\hat{x}_t$.
(c) For all $t\in\mathbb{N}$, for all $(x_t,x_{t+1})\in X_t\times X_{t+1}$, $coA_t''(x_t,x_{t+1})\subset B_t''(\hat{x}_t,\hat{x}_{t+1})$.

Then, for all $T\in\mathbb{N}$, $T\ge 2$, there exist $\lambda_0^T\in\mathbb{R}$ and $p_1^T,\dots,p_T^T\in\mathbb{R}^{n*}$ which satisfy the following conditions:

(i) $\lambda_0^T, p_1^T,\dots,p_T^T$ are not simultaneously equal to zero.
(ii) For all $t\in\{0,\dots,T-1\}$, $p_{t+1}^T\ge 0$ and $\langle p_{t+1}^T, f_t(\hat{x}_t,\hat{u}_t)-\hat{x}_{t+1}\rangle=0$.
(iii) For all $t\in\{1,\dots,T-1\}$, $p_t^T=p_{t+1}^T\circ D_1f_t(\hat{x}_t,\hat{u}_t)+\lambda_0^T D_1\phi_t(\hat{x}_t,\hat{u}_t)$.
(iv) For all $t\in\{0,\dots,T-1\}$, $H_t(\hat{x}_t,\hat{u}_t,p_{t+1}^T,\lambda_0^T)=\max_{u_t\in U_t} H_t(\hat{x}_t,u_t,p_{t+1}^T,\lambda_0^T)$.

Proof. After Proposition 1.2, for all $T\in\mathbb{N}$, $T\ge 2$, the restriction $(\hat{x}_0,\dots,\hat{x}_T,\hat{u}_0,\dots,\hat{u}_{T-1})$ is a solution of $(\mathcal{F}^i(T,\eta,\hat{x}(T)))$. Now we translate $(\mathcal{F}^i(T,\eta,\hat{x}(T)))$ into the form of $(\mathcal{PP})$. We set $\mathfrak{X}:=\prod_{t=1}^{T-1}X_t\subset\mathbb{R}^{(T-1)n}$, and so $N=(T-1)n$, $\mathfrak{U}:=\prod_{t=0}^{T-1}U_t\subset\mathbb{R}^{Tn}$, and so $M=Tn$. We set

$$\begin{aligned}
\Gamma^0(x_1,\dots,x_{T-1},u,\dots,u_{T-1}) &:= \sum_{t=0}^{T-1}\phi_t(x_t,u_t)\\
\Gamma^1(x_1,\dots,x_{T-1},u,\dots,u_{T-1}) &:= f_0^1(\eta,u_0)-x_1^1\\
\Gamma^2(x_1,\dots,x_{T-1},u,\dots,u_{T-1}) &:= f_0^2(x_1,u_1)-x_1^2\\
\dots &:= \dots\\
\Gamma^{Tn}(x_1,\dots,x_{T-1},u,\dots,u_{T-1}) &:= f_{T-1}^n(\eta,u_0)-\hat{x}_{T-1}^n.
\end{aligned}$$

And so, in our case, we have $m_0=m=Tn$, and we have not any equality constraints.

Since the sets X_t are convex, $\mathfrak{X}$ is nonempty and convex as a product of nonempty convex sets, $\mathfrak{U}$ is nonempty as a product of nonempty sets, and $(\hat{x}_1,\dots,\hat{x}_{T-1})\in$

int$\mathfrak{X}$ since $\hat{x}_t \in \mathrm{int}X_t$ for all $t \in \{0, \ldots, T-1\}$. And so the assumption (a) of Theorem 1.5 is fulfilled. Our assumption on the continuity and the differentiability of ϕ_t and f_t implies the assumption (b) of Theorem 1.5.

Note that, for all $t \in \{0, \ldots, T-1\}$,

$$\begin{aligned} A(x_1, \ldots, x_{T-1}) = \{(z^0, \ldots, z^{Tn}) \in \mathbb{R}^{1+Tn} : \exists (u_0, \ldots, u_{T-1}) \in \mathfrak{U} \text{ s.t.} \\ z^0 \le \sum_{t=0}^{T-1} \phi_t(x_t, u_t) \\ z^1 \le f_0^1(\eta, u_0) - x_1^1 \\ \ldots\ldots \\ z^{Tn} \le f_{T-1}^n(\eta, u_0) - x_{T-1}^n\} \end{aligned}$$

and

$$\begin{aligned} B(x_1, \ldots, x_{T-1}) = \{(z^0, \ldots, z^{Tn}) \in \mathbb{R}^{1+Tn} : \exists (u_0, \ldots, u_{T-1}) \in \mathfrak{U}, \\ \exists (v^0, \ldots, v^{Tn}) \in \mathbb{R}^{1+Tn} \text{ s.t.} \\ v^0 z^0 \le \sum_{t=0}^{T-1} \phi_t(x_t, u_t) \\ v^1 z^1 \le f_0^1(\eta, u_0) - x_1^1 \\ \ldots\ldots \\ v^{Tn} z^{Tn} \le f_{T-1}^n(\eta, u_0) - x_{T-1}^n\} \end{aligned}$$

We also introduce the following sets:

$$\begin{aligned} A^0(x_t, x_{t+1}) := \{(z^0, \ldots, z^{Tn}) \in \mathbb{R}^{1+Tn} : \\ \forall k \notin \{0\} \cup \{tn+1, \ldots, (t+1)n\}, z^k = 0, \\ (z^0, z^{tn+1}, \ldots, z^{(t+1)n}) \in A_t''(x_t, x_{t+1})\} \end{aligned}$$

and

$$\begin{aligned} B^0(x_t, x_{t+1}) := \{(z^0, \ldots, z^{Tn}) \in \mathbb{R}^{1+Tn} : \\ \forall k \notin \{0\} \cup \{tn+1, \ldots, (t+1)n\}, z^k = 0, \\ (z^0, z^{tn+1}, \ldots, z^{(t+1)n}) \in B_t''(x_t, x_{t+1})\}. \end{aligned}$$

Note that we have

$$\begin{aligned} \mathrm{co}A^0(x_t, x_{t+1}) := \{(z^0, \ldots, z^{Tn}) \in \mathbb{R}^{1+Tn} : \\ \forall k \notin \{0\} \cup \{tn+1, \ldots, (t+1)n\}, z^k = 0, \\ (z^0, z^{tn+1}, \ldots, z^{(t+1)n}) \in \mathrm{co}A_t''(x_t, x_{t+1})\}. \end{aligned}$$

And so the condition $\mathrm{co}A_t''(x_t, x_{t+1}) \subset B_t''(x_t, x_{t+1})$ implies $\mathrm{co}A_t^0(x_t, x_{t+1}) \subset B_t^0(x_t, x_{t+1})$. Also note that we have the following relations:

$$A(x_1,\ldots,x_{T-1}) = \sum_{t=0}^{T-1} \mathrm{co}A^0(x_t,x_{t+1}),\ B(x_1,\ldots,x_{T-1}) = \sum_{t=0}^{T-1} \mathrm{co}B^0(x_t,x_{t+1}).$$

Then

$$\begin{aligned}\mathrm{co}A(x_1,\ldots,x_{T-1}) &\subset \sum_{t=0}^{T-1} \mathrm{co}A^0(x_t,x_{t+1})\\ \subset \sum_{t=0}^{T-1} \mathrm{co}B^0(x_t,x_{t+1}) &= B(x_1,\ldots,x_{T-1}).\end{aligned}$$

And so the assumption (c) of Theorem 1.5 is fulfilled.

Now we can apply Theorem 1.5 and obtain these conclusions. We set $\lambda_0^T := a_0$, $p_1^T := \sum_{k=1}^n a_k e_k^*$, $p_2^T := \sum_{k=1}^n a_{n+k} e_k^*$, ..., $p_T^T := \sum_{k=1}^n a_{(T-1)n+k} e_k^*$. And then it suffices to translate the conclusions of Theorem 1.5 to obtain our conclusions. Since the $\hat{x}_t$ are interior points, the conclusion (iv) of Theorem 1.5 becomes

$$\frac{\partial}{\partial x^k} \sum_{i=1}^m a_i D_1 \Gamma^i(\hat{x}_1,\ldots,\hat{x}_{T-1},\hat{u}_0,\ldots,\hat{u}_{T-1}) = 0,$$

for all $k \in \{1,\ldots,Tn\}$, which implies the adjoint equation. □

1.4.5 Contributions of Pschenichnyi, Ioffe, and Tihomirov

In [54] (p. 280) the following problem is considered (we have just replaced a minimization by a maximization):

$$(\mathcal{IT})\begin{cases}\text{Maximize } \sum_{i=0}^{T-1} \phi_t(x_t,u_t)\\ \quad\text{when} \quad \forall t \in \{0,\ldots,T-1\},\ x_t \in \mathbb{R}^n, u_t \in U_t\\ \qquad\qquad \forall t \in \{0,\ldots,T-1\},\ x_{t+1} = f_t(x_t,u_t)\\ \qquad\qquad h_0(x_0) = 0,\ h_T(x_T) = 0\\ \qquad\qquad \forall t \in \{0,\ldots,T-1\},\ g_t(x_t) \geq 0\end{cases}$$

where U_t is a nonempty subset of $\mathbb{R}^d$, $\phi_t : \mathbb{R}^n \times U_t \to \mathbb{R}$, $f_t : \mathbb{R}^n \times U_t \to \mathbb{R}^n$, $h_0 : \mathbb{R}^n \to \mathbb{R}^{s_0}$, $h_T : \mathbb{R}^n \to \mathbb{R}^{s_T}$, and $g_t : \mathbb{R}^n \to \mathbb{R}$. Ioffe and Tihomirov establish the following result that they call "discrete maximum principle" saying (in p. 283 of their book) that such a result is present in the book of Pschenichnyi [77].

Theorem 1.7. *Let $(\hat{x}_0,\ldots,\hat{x}_T,\hat{u}_0,,\hat{x}_{T-1})$ be a solution of $(\mathcal{IT})$. We assume that the following conditions are fulfilled:*

(a) $h_0 \in C^1(\mathbb{R}^n,\mathbb{R}^{s_0})$ and $h_T \in C^1(\mathbb{R}^n,\mathbb{R}^{s_T})$.

(b) *For all* $t \in \{0, \ldots, T-1\}$, $\phi_t \in C^0(\mathbb{R}^n \times U_t, \mathbb{R})$ *and, for all* $(x, u) \in \mathbb{R}^n \times U_t$, *the partial differential* $D_1\phi_t(x, u)$ *exists and* $D_1\phi_t \in C^0(\mathbb{R}^n \times U_t, \mathbb{R}^{n*})$.

(c) *For all* $t \in \{0, \ldots, T-1\}$, $f_t \in C^0(\mathbb{R}^n \times U_t, \mathbb{R}^n)$ *and, for all* $(x, u) \in \mathbb{R}^n \times U_t$, *the partial differential* $D_1 f_t(x, u)$ *exists and* $D_1\phi_t \in C^0(\mathbb{R}^n \times U_t, \mathcal{L}(\mathbb{R}^n, \mathbb{R}^n))$.

(d) *For all* $t \in \{0, \ldots, T-1\}$, *for all* $x \in \mathbb{R}^n$, *for all* $u, v \in U_t$, *for all* $r \in [0, 1]$, *there exists* $w \in U_t$ *such that*

$$\begin{cases} \phi_t(x, w) \geq (1-r)\phi_t(x, u) + r\phi_t(x, v) \\ f_t(x, w) = (1-r) f_t(x, u) + r f_t(x, v). \end{cases}$$

Then there exist $\lambda_0 \geq 0$, $\mu_1, \ldots, \mu_{T-1} \in \mathbb{R}$, $p_0, \ldots, p_T \in \mathbb{R}^{n*}$, $\ell_0 \in \mathbb{R}^{s_0*}$, $\ell_T \in \mathbb{R}^{s_T*}$ *which satisfy the following conditions:*

(i) $\lambda_0, \mu_1, \ldots, \mu_{T-1}, p_0, \ldots, p_T, \ell_0, \ell_T$ *are not simultaneously equal to zero.*

(ii) $p_t = D_1 H_t(\hat{x}_t, \hat{u}_t, p_{t+1}, \lambda_0) - \mu_t Dg_t(\hat{x}_t)$ *for all* $t \in \{1, \ldots, T-1\}$.

(iii) $p_0 = \ell_0 \circ Dh_0(\hat{x}_0)$ *and* $p_T = -\ell_T \circ Dh_T(\hat{x}_T)$.

(iv) $H_t(\hat{x}_t, \hat{u}_t, p_{t+1}, \lambda_0) = \max_{u \in U_t} H_t(\hat{x}_t, u, p_{t+1}, \lambda_0)$ *for all* $t \in \{0, \ldots, T-1\}$.

(v) $\mu_t g_t(\hat{x}_t) = 0$ *for all* $t \in \{1, \ldots, T-1\}$.

Now we want to apply this theorem to our problems $(\mathcal{F}^e(T, \eta, \hat{x}_T))$. We have $s_0 = s_T = n$ and $h_0(x_0) = x_0 - \eta = 0$, $h_T(x_T) = x_T - \hat{x}_T = 0$. Since the constraints $g_t \geq 0$ are not present in $(\mathcal{F}^e(T, \eta, \hat{x}_T))$, the associated multipliers are not present. And so we straightforwardly obtain the following result.

Proposition 1.9. *Let* $(\underline{\hat{x}}, \underline{\hat{u}})$ *be a solution of* $(\mathcal{P}_e^n)$, *or of* $(\mathcal{P}_e^s)$, *or of* $(\mathcal{P}_e^o)$, *or of* $(\mathcal{P}_e^w)$. *We assume that the following conditions are fulfilled:*

(a) *For all* $t \in \mathbb{N}$, X_t *is a nonempty open subset of* $\mathbb{R}^n$, *and* U_t *is a nonempty subset of* $\mathbb{R}^d$.

(b) *For all* $t \in \mathbb{N}$, $\phi_t \in C^0(X_t \times U_t, \mathbb{R})$ *and, for all* $(x, u) \in \mathbb{R}^n \times U_t$, *the partial differential* $D_1\phi_t(x, u)$ *exists and* $D_1\phi_t \in C^0(X_t \times U_t, \mathbb{R}^{n*})$.

(c) *For all* $t \in \mathbb{N}$, $f_t \in C^0(X_t \times U_t, \mathbb{R}^n)$ *and, for all* $(x, u) \in \mathbb{R}^n \times U_t$, *the partial differential* $D_1 f_t(x, u)$ *exists and* $D_1\phi_t \in C^0(X_t \times U_t, \mathcal{L}(\mathbb{R}^n, \mathbb{R}^n))$.

(d) *For all* $t \in \mathbb{N}$, *for all* $x \in X_t$, *for all* $u, v \in U_t$, *for all* $r \in [0, 1]$, *there exists* $w \in U_t$ *such that*

$$\begin{cases} \phi_t(x, w) \geq (1-r)\phi_t(x, u) + r\phi_t(x, v) \\ f_t(x, w) = (1-r) f_t(x, u) + r f_t(x, v). \end{cases}$$

Then, for all $T \in \mathbb{N}$, $T \geq 2$, *there exist* $\lambda_0^T \in \mathbb{R}_+$, $p_1^T, \ldots, p_T^T \in \mathbb{R}^{n*}$ *which satisfy the following conditions:*

(i) $\lambda_0^T, p_1^T, \ldots, p_T^T$ *are not simultaneously equal to zero.*

(ii) $p_t^T = D_1 H_t(\hat{x}_t, \hat{u}_t, p_{t+1}, \lambda_0^T)$ *for all* $t \in \{1, \ldots, T-1\}$.

(iii) $H_t(\hat{x}_t, \hat{u}_t, p_{t+1}, \lambda_0^T) = \max_{u \in U_t} H_t(\hat{x}_t, u, p_{t+1}, \lambda_0^T)$ *for all* $t \in \{0, \ldots, T-1\}$.

The previous result concerns the systems which are governed by (DE). To obtain a similar result for systems which are governed by (DI), we use a result on a parametrized static optimization problem. This result is due to Ioffe and Tihomirov (Theorem 3, p. 71, in [54]). These authors present it as an improvement of a result of Pschenichnyi and Nenahov. They establish it in the setting of Banach spaces. In this chapter we will use it only in finite-dimensional normed spaces. Their problem is the following one.

$$(\mathscr{P}\mathscr{P}2)\begin{cases} \text{Maximize } \Gamma^0(x,u) \\ \quad \text{when} \quad F(x,u) = 0 \\ \qquad\qquad \forall j \in \{1,\dots,m\}, \ \ \Gamma^j(x,u) \geq 0 \end{cases}$$

where X, Y are Banach spaces, U is a nonempty set, $\Gamma^j : X \times U \to \mathbb{R}$ for all $j \in \{1,\dots,m\}$, and $F : X \times U \to Y$. The generalized Lagrangian of this problem is $\mathscr{G} : X \times U \times \mathbb{R}^{1+m} \times Y^* \to \mathbb{R}$ defined by

$$\mathscr{G}(x,u,\lambda_0,\lambda_1,\dots,\lambda_m,\pi) := \sum_{j=0}^{m} \lambda_j \Gamma^j(x,u) + \langle \pi, F(x,u)\rangle.$$

Theorem 1.8. *Let $(\overline{x},\overline{u})$ be a solution of $(\mathscr{P}\mathscr{P}2)$. We assume that the following conditions are fulfilled:*

(a) There exists a neighborhood V of $\overline{x}$ in X such that

(a1) For all $u \in U$, $F(.,u)$ and the $\Gamma^j(.,u)$ are of class C^1 at $\overline{x}$.
(a2) For all $x \in V$, for all $u_1,u_2 \in U$, for all $\theta \in [0,1]$, there exists $u_3 \in U$ satisfying

$$\begin{aligned} F(x,u_3) &= (1-\theta)F(x,u_1) + \theta F(x,u_2) \\ \Gamma^j(x,u_3) &\geq (1-\theta)\Gamma^j(x,u_1) + \theta\Gamma^j(x,u_2) \end{aligned}$$

for all $j \in \{0,\dots,m\}$.

(b) The codimension of $D_1F(\overline{x},\overline{u})$ is finite.

Then there exist $(\lambda_j)_{0\leq j\leq m} \in \mathbb{R}^{1+m}$ and $\pi \in Y^*$ which satisfy the following conditions:

(i) $(\lambda_j)_{0\leq j\leq m}$ and π are not simultaneously equal to zero.
(ii) For all $j \in \{0,\dots,m\}$, $\lambda_j \geq 0$.
(iii) For all $j \in \{1,\dots,m\}$, $\lambda_j\Gamma^j(\overline{x},\overline{u}) = 0$.
(iv) $D_1\mathscr{G}(\overline{x},\overline{u},\lambda_0,\lambda_1,\dots,\lambda_m,\pi) = 0$.
(v) $\mathscr{G}(\overline{x},\overline{u},\lambda_0,\lambda_1,\dots,\lambda_m,\pi) = \max_{u\in U} \mathscr{G}(\overline{x},u,\lambda_0,\lambda_1,\dots,\lambda_m,\pi)$.

Now we want to use this theorem on our problem $(\mathscr{F}_i(T,\eta,\hat{x}(T)))$. Note that in the previous result we can replace X by an open nonempty subset of X.

Proposition 1.10. *Let $(\underline{\hat{x}}, \underline{\hat{u}})$ be a solution of $(\mathscr{P}_i^n)$, or of $(\mathscr{P}_i^s)$, or of $(\mathscr{P}_i^o)$, or of $(\mathscr{P}_i^w)$. We assume that the following conditions are fulfilled;*

(a) For all $t \in \mathbb{N}$, $\phi_t(., \hat{u}_t)$ and $f_t(., \hat{u}_t)$ are of class C^1 at $\hat{x}$.
(b) For all $t \in \mathbb{N}$, there exists a neighborhood V_t of $\hat{x}_t$ in X_t such that, for all $x \in V_t$, for all $u_1, u_2 \in U_t$, for all $\theta \in [0, 1]$, there exists $u_3 \in U_t$ such that

$$\begin{cases} \phi_t(x, u_3) \geq (1-\theta)\phi_t(x, u_1) + \theta\phi_t(x, u_2) \\ f_t(x, u_3) \geq (1-\theta) f_t(x, u_1) + \theta f_t(x, u_2). \end{cases}$$

Then, for all $T \in \mathbb{N}$, $T \geq 2$, there exist $\lambda_0^T \in \mathbb{R}$ and $p_1^T, \ldots, p_T^T \in \mathbb{R}^{n*}$ which satisfy the following conditions:

(i) λ_0^T and $p_1^T, \ldots, p_T^T$ are not simultaneously equal to zero.
(ii) $\lambda_0^T \geq 0$.
(iii) $p_{t+1}^T \geq 0$ and $\langle p_{t+1}^T, f_t(\hat{x}_t, \hat{u}_t) - \hat{x}_{t+1}\rangle = 0$, for all $t \in \{0, \ldots, T-1\}$.
(iv) $p_t^T = D_1 H_t(\hat{x}_t, \hat{u}_t, p_{t+1}^T, \lambda_0^T)$ for all $t \in \{1, \ldots, T-1\}$.
(v) $H_t(\hat{x}_t, \hat{u}_t, p_{t+1}^T, \lambda_0^T) = \max_{u_t \in U_t} H_t(\hat{x}_t, u_t, p_{t+1}^T, \lambda_0^T)$ for all $t \in \{0, \ldots T-1\}$.

Proof. Using Proposition 1.2, the restriction $(\hat{x}_0, \ldots, \hat{x}_T, \hat{u}_0, \ldots, \hat{u}_{T-1})$ is a solution of $(\mathscr{F}_i(T, \eta, \hat{x}_T))$, and consequently $(\hat{x}_1, \ldots, \hat{x}_{T-1}, \hat{u}_0, \ldots, \hat{u}_{T-1})$ is a solution of a problem as $(\mathscr{P}\mathscr{P}2)$ without the presence of F since we do not have equality constraints. And so we set $X = \prod_{t=1}^{T-1} X_t$, $U = \prod_{t=0}^{T-1} U_t$,

$$\begin{aligned}
\Gamma^0(x_1, \ldots, x_{T-1}, u_0, \ldots, u_{T-1}) &:= \sum_{t=0}^{T-1} \phi_t(x_t, u_t) \\
\Gamma^1(x_1, \ldots, x_{T-1}, u_0, \ldots, u_{T-1}) &:= f_0^1(\eta, u_0) - x_1^1 \\
\Gamma^2(x_1, \ldots, x_{T-1}, u_0, \ldots, u_{T-1}) &:= f_0^2(\eta, u_0) - x_1^2 \\
\cdots\cdots\cdots\cdots & \cdots\cdots\cdots \\
\Gamma^{Tn}(x_1, \ldots, x_{T-1}, u_0, \ldots, u_{T-1}) &:= f_{T-1}^n(x_{T-1}, u_{T-1}) - \hat{x}_T^n.
\end{aligned}$$

Our assumptions (a) and (b) imply that the assumptions of Theorem 1.8 are fulfilled. And so we can use Theorem 1.8 and we can assert that there exist λ_0^T, $\lambda_1^T, \ldots, \lambda_{Tn}^T \in \mathbb{R}$ such that the following conditions hold:

(i') $(\lambda_j^T)_{0 \leq j \leq Tn}$ is not equal to zero.
(ii') For all $j \in \{0, \ldots, Tn\}$, $\lambda_j^T \geq 0$.
(iii') For all $j \in \{1, \ldots, Tn\}$, $\lambda_j^T \Gamma^j(\hat{x}_1, \ldots, \hat{u}_{T-1}) = 0$.
(iv') For all $t \in \{1, \ldots, T-1\}$, $D_{x_t}\mathscr{G}(\hat{x}_1, \ldots, \hat{u}_{T-1}, \lambda_0^T, , \lambda_{Tn}^T, 0) = 0$.
(v') $\mathscr{G}(\hat{x}_1, \ldots, \hat{x}_{T-1}, \hat{u}_0, \ldots, \hat{u}_{T-1}, \lambda_0^T, , \hat{u}_0, \ldots, \lambda_{Tn}^T, 0) =$
$\max_{(u_0, \ldots, u_{T-1}) \in U} \mathscr{G}(\hat{x}_1, \ldots, \hat{x}_{T-1}, u_0, \ldots, u_{T-1}, \lambda_0^T, , \hat{u}_0, \ldots, \lambda_{Tn}^T, 0)$.

And then, it suffices to translate these conditions. First we set $p_{t+1}^T := \sum_{j=1}^n \lambda_{j+tn}^T e_j^* \in \mathbb{R}^{n*}$. And so (i') implies our conclusion (i), (ii') implies our conclusion (ii) since $\lambda^T_{j+tn} \geq 0$ and $e^*_j \geq 0$, and (iii') implies $\langle p^T_{t+1}, f_t(\hat{x}_t, \hat{u}_t) - \hat{x}_{t+1}\rangle = 0$ for all $t \in \{0, \ldots, T-1\}$ that is our conclusion (iii). Note that

$$\begin{aligned}
&\mathcal{G}(x_1, \ldots, x_{T-1}, u_0, \ldots, u_{T-1}, \lambda^T_0, \ldots, \lambda^T_{Tn}, 0) \\
&\quad = \lambda^T_0 \sum_{t=0}^{T-1} \phi_t(x_t, u_t) + \sum_{t=0}^{T-1} \langle p^T_{t+1}, f_t(x_t, u_t) - x_{t+1}\rangle \\
&\quad = \sum_{t=0}^{T-1} H_t(x_t, u_t, p^T_{t+1}, \lambda^T_0) - \sum_{t=0}^{T-1} \langle p^T_{t+1}, x_{t+1}\rangle,
\end{aligned}$$

where $x_0 = \hat{x}_0$ and $x_T = \hat{x}_T$. Then (iv') implies, for all $t \in \{1, \ldots, T-1\}$, that

$$\begin{aligned}
0 &= D_{x_t}\mathcal{G}(\hat{x}_1, \ldots, \hat{x}_t, \ldots, \hat{u}_{T-1}, \lambda^T_0, , \lambda^T_{Tn}, 0) \\
&= D_1 H_t(\hat{x}_t, \hat{u}_t, p^T_{t+1}, \lambda^T_0) - p^T_t,
\end{aligned}$$

that is our conclusion (iv). Lastly, from (v'), we deduce that, for all $t \in \{0, \ldots, T-1\}$,

$$\begin{aligned}
&\mathcal{G}(\hat{x}_1, \ldots, \hat{u}_0, \ldots, \hat{u}_t, \ldots, \hat{u}_{T-1}, \lambda^T_0, , \hat{u}_0, \ldots, \lambda^T_{Tn}, 0) \\
&\quad = \max_{u_t \in U_t} \mathcal{G}(\hat{x}_1, \ldots, \hat{u}_0, \ldots, u_t, \ldots, \hat{u}_{T-1}, \lambda^T_0, , \hat{u}_0, \ldots, \lambda^T_{Tn}, 0),
\end{aligned}$$

and by using the previous formula which relies $\mathcal{G}$ and the H_t, we deduce

$$\begin{aligned}
&\sum_{s=0}^{T-1} H_s(\hat{x}_s, \hat{u}_s, p^T_{s+1}, \lambda^T_0) - \sum_{s=0}^{T-1} \langle p^T_{s+1}, \hat{x}_{s+1}\rangle \\
&\quad = \max_{u_t \in U_t} \Big(\sum_{s=0}^{T-1} H_s(\hat{x}_s, u_s, p^T_{s+1}, \lambda^T_0) - \sum_{t=0}^{T-1} \langle p^T_{s+1}, \hat{x}_{s+1}\rangle\Big),
\end{aligned}$$

for all $t \in \{0, .., T-1\}$ and deleting all the terms where u_t does not appear, we obtain our conclusion (v). □

Chapter 2
Infinite-Horizon Theorems

2.1 Introduction

In this chapter, we give infinite-horizon theorems using the tools of the previous chapter. In Sect. 2.2 we present several weak maximum principles which are obtained through the method of reduction to finite horizon. We successively use two additional conditions to obtain results in the infinite-horizon setting from results of the finite-horizon setting. In Sect. 2.3 we present several strong maximum principles which are obtained through the method of reduction to finite horizon. We successively use three additional conditions which permit the extension of finite-horizon results into infinite-horizon results. In Sect. 2.4 we study constrained problems and in Sect. 2.5 multiobjective problems.

2.2 Weak Pontryagin Principles in Infinite Horizon

The first establishment of a Pontryagin principle in infinite horizon in the framework of the continuous time is due to Halkin [34]. The major difficulty to adapt the proof of Halkin to the discrete-time framework is the following one: whereas integrating an ordinary differential equation forward or integrating it backward is the same thing, it is not the same for a difference equation. And the so-called *adjoint equations*, $p_t = D_1 H_t(\hat{x}_t, \hat{u}_t, p_{t+1}, \lambda_0)$, are backward difference equations. To overcome this difficulty, we propose several solutions. Each of the following subsections provides a solution to this difficulty.

J. Blot and N. Hayek, *Infinite-Horizon Optimal Control in the Discrete-Time Framework*, SpringerBriefs in Optimization, DOI 10.1007/978-1-4614-9038-8_2,

2.2.1 A Condition of Invertibility

This condition concerns the partial differential with respect to the state variable of the vector field of the equation of motion. It is the following: the invertibility of $D_1 f_t(\hat{x}_t, \hat{u}_t)$. This invertibility allows to transform the adjoint equation into a forward difference equation. In this context, this condition appears for the first time in the paper of Blot and Chebbi [16].

We need two lemmas to prove the theorems of this section. First, the following lemma is a way to express the Diagonal Process of Cantor.

Lemma 2.1. *Let Z be a real finite-dimensional normed vector space. For all $(t, T) \in \mathbb{N}_* \times \mathbb{N}_*$ such that $t \leq T$ we consider $z_t^T \in Z$. We assume that, for all $t \in \mathbb{N}_*$, the sequence $T \mapsto z_t^T$ is bounded. Then there exists an increasing function $\sigma : \mathbb{N}_* \to \mathbb{N}_*$ such that, for all $t \in \mathbb{N}_* \to \mathbb{N}_*$ such that, for all $t \in \mathbb{N}_*$, there exists $z_t \in \mathbb{N}_*$.*

A precise proof of this lemma is given in Theorem A.1 in Appendix A. The second lemma expresses a relation between a property related to the norm and an algebraic property.

Lemma 2.2. *Let Z be a real finite-dimensional normed vector space. Let $z_1, \ldots, z_k$ be linearly independent vectors in Z, and for all $j \in \{1, \ldots, k\}$, let $(r_t^j)_{t \in \mathbb{N}}$ be a real sequence. If the vector sequence $\left(\sum_{j=1}^{k} r_t^j v_j \right)_{t \in \mathbb{N}}$ is bounded in Z, then the real sequence $(r_t^j)_{t \in \mathbb{N}}$ is bounded in $\mathbb{R}$ for all $j \in \{1, \ldots, k\}$.*

A precise proof of this lemma is given in [13] p. 48.

First we use the multiplier rule of Halkin to obtain the following result.

Theorem 2.1. *Let $(\underline{\hat{x}}, \underline{\hat{u}})$ be a solution of $(\mathcal{P}_a^n)$, or of $(\mathcal{P}_a^s)$, or of $(\mathcal{P}_a^o)$, or of $(\mathcal{P}_a^w)$ for $a \in \{e, i\}$. We assume that the following conditions are fulfilled:*

(a) For all $t \in \mathbb{N}$, X_t is a nonempty open subset of $\mathbb{R}^n$.
(b) For all $t \in \mathbb{N}$,

$$U_t = \left(\bigcap_{\alpha=1}^{k^i} \{u \in \mathbb{R}^d : g_t^\alpha(u) \geq 0\} \right) \cap \left(\bigcap_{\beta=1}^{k^e} \{u \in \mathbb{R}^d : h_t^\beta(u) = 0\} \right)$$

where $g_t^\alpha : \mathbb{R}^d \to \mathbb{R}$ and $h_t^\beta : \mathbb{R}^d \to \mathbb{R}$ are continuous on a neighborhood of $\hat{u}_t$ and they are differentiable at $\hat{u}_t$, for all $\alpha \in \{1, \ldots, k^i\}$ and for all $\beta \in \{1, \ldots, k^e\}$, and we assume that $U_t \neq \emptyset$.
(c) For all $t \in \mathbb{N}$, the differentials $Dg_t^1(\hat{u}_t), \ldots, Dg_t^{k^i}(\hat{u}_t)$, $Dh_t^1(\hat{u}_t), \ldots, Dh_t^{k^e}(\hat{u}_t)$ are linearly independent.
(d) For all $t \in \mathbb{N}$, ϕ_t and f_t are continuous on a neighborhood of $(\hat{x}_t, \hat{u}_t)$ and they are differentiable at $(\hat{x}_t, \hat{u}_t)$.
(e) For all $t \in \mathbb{N}$, the partial differential $D_1 f_t(\hat{x}_t, \hat{u}_t)$ is invertible.

Then there exist $\lambda_0 \in \mathbb{R}$, $(p_t)_{t\in\mathbb{N}_*} \in (\mathbb{R}^{n*})^{\mathbb{N}_*}$, $(\lambda_{1,t})_{t\in\mathbb{N}} \in \mathbb{R}^{\mathbb{N}}, \dots, (\lambda_{k^i,t})_{t\in\mathbb{N}} \in \mathbb{R}^{\mathbb{N}}$ $(\mu_{1,t})_{t\in\mathbb{N}} \in \mathbb{R}^{\mathbb{N}}, \dots, (\mu_{k^e,t})_{t\in\mathbb{N}} \in \mathbb{R}^{\mathbb{N}}$ *which satisfy the following conditions:*

(i) $(\lambda_0, p_1) \neq (0,0)$.
(ii) $\lambda_0 \geq 0$.
(iii) For all $t \in \mathbb{N}$, $p_{t+1} \geq 0$ *and* $\langle p_{t+1}, f_t(\hat{x}_t, \hat{u}_t) - \hat{x}_{t+1}\rangle = 0$ *when* $a = i$.
(iv) For all $t \in \mathbb{N}$, *for all* $\alpha \in \{1, \dots, k^i\}$, $\lambda_{\alpha,t} \geq 0$.
(v) For all $t \in \mathbb{N}$, *for all* $\alpha \in \{1, \dots, k^i\}$, $\lambda_{\alpha,t} g_t^\alpha(\hat{u}_t) = 0$.
(vi) For all $t \in \mathbb{N}_*$, $p_t = p_{t+1} \circ D_1 f_t(\hat{x}_t, \hat{u}_t) + \lambda_0 D_1\phi_t(\hat{x}_t, \hat{u}_t)$.
(vii) For all $t \in \mathbb{N}$,

$$D_2 H_t(\hat{x}_t, \hat{u}_t, p_{t+1}, \lambda_0) + \sum_{\alpha=1}^{k^i} \lambda_{\alpha,t} D g_t^\alpha(\hat{u}_t) + \sum_{\beta=1}^{k^e} \mu_{\beta,t} D h_t^\beta(\hat{u}_t) = 0.$$

Proof. Using Proposition 1.2 we know that for all $T \in \mathbb{N}$, $T \geq 2$, the restriction $(\hat{x}_0, \dots, \hat{x}_{T-1}, \hat{u}_0, \dots, \hat{u}_{T-1})$ is a solution of $(\mathcal{F}^a(T, \eta, \hat{x}_T))$ for $a \in \{e, i\}$. And consequently the conclusions of Proposition 1.3 hold when $a = e$, and the conclusions of Proposition 1.4 hold when $a = i$. Using the assumption (e), the conclusion (iv) of Proposition 1.3 or of Proposition 1.4 which is

$$p_t^T = p_{t+1}^T \circ D_1 f_t(\hat{x}_t, \hat{u}_t) + \lambda_0^T D_1\phi_t(\hat{x}_t, \hat{u}_t),$$

for all $t \in \{0, \dots, T-1\}$, becomes

$$p_{t+1}^T = p_t^T \circ (D_1 f_t(\hat{x}_t, \hat{u}_t))^{-1} - \lambda_0^T D_1\phi_t(\hat{x}_t, \hat{u}_t) \circ (D_1 f_t(\hat{x}_t, \hat{u}_t))^{-1}. \tag{2.1}$$

Using (2.1), we see that $(\lambda_0^T, p_1^T) = (0,0)$ implies $(\lambda_0^T, p_1^T, \dots, p_T^T) = (0, 0, \dots, 0)$. And so by contraposition we obtain the following relation.

$$(\lambda_0^T, p_1^T, \dots, p_T^T) \neq (0, 0, \dots, 0) \Longrightarrow (\lambda_0^T, p_1^T) \neq (0,0). \tag{2.2}$$

Using assumption (c) and the last assertion of Proposition 1.3 or of Proposition 1.4 we obtain $(\lambda_0^T, p_1^T) \neq (0,0)$. Since the set of all lists of multipliers is a cone, we can normalize these lists of multipliers, and so we can choose:

$$\forall T \in \mathbb{N}_*, \ \ \|(\lambda_0^T, p_1^T)\| = 1. \tag{2.3}$$

From (2.1) the sequence $T \mapsto p_1^T$ is bounded in $\mathbb{R}^{n*}$, and the sequence $T \mapsto \lambda_0^T$ is bounded in $\mathbb{R}$. From (2.1) we obtain

$$\|p_2^T\| \leq \|p_1^T\|.\|(D_1 f_1(\hat{x}_1, \hat{u}_1))^{-1}\| + |\lambda_0^T|.\|D_1\phi_1(\hat{x}_1, \hat{u}_1)\|.\|(D_1 f_1(\hat{x}_1, \hat{u}_1))^{-1}\|$$

which implies

$$\sup_{T\geq 2} \|p_2^T\| \leq \|(D_1 f_1(\hat{x}_1, \hat{u}_1))^{-1}\| + \|D_1\phi_1(\hat{x}_1, \hat{u}_1)\|.\|(D_1 f_1(\hat{x}_1, \hat{u}_1))^{-1}\| < +\infty,$$

and proceeding by induction, we obtain

$$\sup_{T \geq t+1} \|p_{t+1}^T\| \leq \sup_{T \geq t} \|p_t^T\|.\|(D_1 f_t(\hat{x}_t, \hat{u}_t))^{-1}\| + \|D_1 \phi_t(\hat{x}_t, \hat{u}_t)\|.\|(D_1 f_t(\hat{x}_t, \hat{u}_t))^{-1}\|$$

which implies

$$\forall t \in \mathbb{N}, \quad \sup_{T \geq t} \|p_t^T\| < +\infty. \tag{2.4}$$

From (2.2) we deduce that the sequence $T \mapsto D_2 H_t(\hat{x}_t, \hat{u}_t, p_{t+1}^T, \lambda_0^T)$ is bounded and using the conclusion (v) of Proposition 1.3 or of Proposition 1.4 we obtain that the sequence $T \mapsto \sum_{\alpha=1}^{k^i} \lambda_{\alpha,t}^T Dg_t^\alpha(\hat{u}_t) + \sum_{\beta=1}^{k^e} \mu_{\beta,t}^T Dh_t^\beta(\hat{u}_t)$ is bounded, and by using the assumption (c) and Lemma 2.2 we obtain the following relations:

$$\forall t \in \mathbb{N}, \quad \sup_{T \geq t} |\lambda_{\alpha,t}^T| < +\infty. \tag{2.5}$$

$$\forall t \in \mathbb{N}, \quad \sup_{T \geq t} |\mu_{\beta,t}^T| < +\infty. \tag{2.6}$$

After (2.1)–(2.4), using Lemma 2.1 we know that there exist an increasing function $\sigma : \mathbb{N}_* \to \mathbb{N}_*$, $\lambda_0 \in \mathbb{R}$, $p_t \in \mathbb{R}^{n*}$, $\lambda_{\alpha,t} \in \mathbb{R}$, $\mu_{\beta,t} \in \mathbb{R}$ for all $t \in \mathbb{N}$, for all $\alpha \in \{1, \ldots, k^i\}$, for all $\beta \in \{1, \ldots, k^e\}$ such that, for all $t \in \mathbb{N}$, for all $\alpha \in \{1, \ldots, k^i\}$, for all $\beta \in \{1, \ldots, k^e\}$, the following equalities hold:

$$\left.\begin{aligned} \lim_{T \to +\infty} \lambda_0^{\sigma(T)} &= \lambda_0 \\ \lim_{T \to +\infty} p_{t+1}^{\sigma(T)} &= p_{t+1} \\ \lim_{T \to +\infty} \lambda_{\alpha,t}^{\sigma(T)} &= \lambda_{\alpha,t} \\ \lim_{T \to +\infty} \mu_{\beta,t}^{\sigma(T)} &= \mu_{\beta,t}. \end{aligned}\right\} \tag{2.7}$$

From (2.3), (2.7), and the continuity of the norm we obtain $\|(\lambda_0, p_1)\| = 1$, which implies the conclusion (i). From (2.7) and the conclusions of Proposition 1.3 or of Proposition 1.4, we obtain the conclusions (ii), (iii), (iv), and (v) by taking $T \to +\infty$. □

When we have $\hat{u}_t \in \operatorname{int} U_t$ for all $t \in \mathbb{N}$, using Corollary 1.1 instead of Proposition 1.3 and Corollary 1.2 instead of Proposition 1.4, a similar reasoning allows to establish the following result.

Theorem 2.2. *Let* $(\underline{\hat{x}}, \underline{\hat{u}})$ *be a solution of* $(\mathcal{P}_a^n)$, *or of* $(\mathcal{P}_a^s)$, *or of* $(\mathcal{P}_a^o)$, *or of* $(\mathcal{P}_a^w)$ *for* $a \in \{e, i\}$. *We assume that the following conditions are fulfilled:*

(a) For all $t \in \mathbb{N}$, X_t *is a nonempty open subset of* $\mathbb{R}^n$.
(b) For all $t \in \mathbb{N}$, $\hat{u}_t \in \operatorname{int} U_t$.

(c) For all $t \in \mathbb{N}$, ϕ_t and f_t are continuous on a neighborhood of $(\hat{x}_t, \hat{u}_t)$ and they are differentiable at $(\hat{x}_t, \hat{u}_t)$.
(d) For all $t \in \mathbb{N}$, the partial differential $D_1 f_t(\hat{x}_t, \hat{u}_t)$ is invertible.

Then there exist $\lambda_0 \in \mathbb{R}$, $(p_t)_{t \in \mathbb{N}_*} \in (\mathbb{R}^{n*})^{\mathbb{N}_*}$ which satisfy the following conditions:

(i) $(\lambda_0, p_1) \neq (0,0)$.
(ii) $\lambda_0 \geq 0$.
(iii) For all $t \in \mathbb{N}$, $p_{t+1} \geq 0$ and $\langle p_{t+1}, f_t(\hat{x}_t, \hat{u}_t) - \hat{x}_{t+1}\rangle = 0$ when $a = i$.
(iv) For all $t \in \mathbb{N}_*$, $p_t = p_{t+1} \circ D_1 f_t(\hat{x}_t, \hat{u}_t) + \lambda_0 D_1 \phi_t(\hat{x}_t, \hat{u}_t)$.
(v) For all $t \in \mathbb{N}$, $\quad D_2 H_t(\hat{x}_t, \hat{u}_t, p_{t+1}, \lambda_0) = 0$.

After the use of the multiplier rule of Halkin, we use the multiplier rule of Clarke.

Theorem 2.3. *Let $(\underline{\hat{x}}, \underline{\hat{u}})$ be a solution of $(\mathscr{P}^n_a)$, or of $(\mathscr{P}^s_a)$, or of $(\mathscr{P}^o_a)$, or of $(\mathscr{P}^w_a)$ for $a \in \{e, i\}$. We assume that the following conditions are fulfilled:*

(a) For all $t \in \mathbb{N}$, ϕ_t is Lipschitzian on a neighborhood of $(\hat{x}_t, \hat{u}_t)$ and regular at $(\hat{x}_t, \hat{u}_t)$.
(b) For all $t \in \mathbb{N}$, f_t is strictly differentiable at $(\hat{x}_t, \hat{u}_t)$.
(c) For all $t \in \mathbb{N}$, U_t is closed and Clarke-regular at $\hat{u}_t$.
(d) For all $t \in \mathbb{N}$, the partial differential $D_1 f_t(\hat{x}_t, \hat{u}_t)$ is invertible.

Then there exist $\lambda_0 \in \mathbb{R}$, $(p_t)_{t \in \mathbb{N}_*} \in (\mathbb{R}^{n*})^{\mathbb{N}_*}$ which satisfy the following conditions:

(i) $(\lambda_0, p_1) \neq (0,0)$.
(ii) $\lambda_0 \geq 0$.
(iii) For all $t \in \mathbb{N}$, $p_{t+1} \geq 0$ and $\langle p_{t+1}, f_t(\hat{x}_t, \hat{u}_t) - \hat{x}_{t+1}\rangle = 0$ when $a = i$.
(iv) For all $t \in \mathbb{N}_*$, $p_t \in \partial_2 H_t(\hat{x}_t, \hat{u}_t, p_{t+1}, \lambda_0)$.
(v) For all $t \in \mathbb{N}$, $\partial_2 H_t(\hat{x}_t, \hat{u}_t, p_{t+1}, \lambda_0) \cap N_{U_t}(\hat{u}_t) \neq \emptyset$, where $N_{U_t}(\hat{u}_t)$ is the normal cone of U_t at $\hat{u}_t$.

Proof. Using Proposition 1.2, the restriction $(\hat{x}_0, \ldots, \hat{x}_{T-1}, \hat{u}_0, \ldots, \hat{u}_{T-1})$ is a solution of $(\mathscr{F}^a(T, \eta, \hat{x}_T))$, for $a \in \{e, i\}$. Consequently using Proposition 1.5 when $a = e$ and Proposition 1.6 when $a = i$, we know that, for all $T \in N$, $T \geq 2$, there exist $\lambda_0^T \in \mathbb{R}$, $p_{t+1}^T \in \mathbb{R}^{n*}$, when $t \in \{0, \ldots, T-1\}$, which satisfy the following conditions:

$$\left(\lambda_0^T, p_1^T, \ldots, p_T^T\right) \neq (0, 0, \ldots, 0) \tag{2.8}$$

$$\lambda_0^T \geq 0 \tag{2.9}$$

$$\forall t \in \{0, \ldots, T-1\}, \exists \varphi_t^T \in \partial_1 \phi_t(\hat{x}_t, \hat{u}_t) \text{ s.t. } p_t^T = \lambda_0^T \varphi_t^T + p_{t+1}^T \circ D_1 f_t(\hat{x}_t, \hat{u}_t) \tag{2.10}$$

$$\left.\begin{array}{r} \forall t \in \{0, \ldots, T-1\}, \exists \psi_t^T \in \partial_2 \phi_t(\hat{x}_t, \hat{u}_t) \text{ s.t.} \\ \forall v_t \in T_{U_t}(\hat{u}_t), \langle \lambda_0^T \psi_t^T + p_{t+1}^T \circ D_2 f_t(\hat{x}_t, \hat{u}_t), v_t\rangle \leq 0 \end{array}\right\} \tag{2.11}$$

Using assumption (d) we can transform (2.10) into the following relation:

$$p_{t+1}^T = p_t^T \circ (D_1 f_t(\hat{x}_t, \hat{u}_t))^{-1} - \lambda_0^T \varphi_t^T \circ (D_1 f_t(\hat{x}_t, \hat{u}_t))^{-1}. \tag{2.12}$$

Reasoning as in the proof of Theorem 2.2, from (2.8) and (2.12) we deduce that $(\lambda_0^T, p_1^T) \neq (0,0)$ and we can choose (λ_0^T, p_1^T) such that $\|(\lambda_0^T, p_1^T)\| = 1$. And so the sequences $T \mapsto \lambda_0^T$ and $T \mapsto p_1^T$ are bounded, and from (2.12) we deduce by induction that, for all $t \in \mathbb{N}$, all the sequences $T \mapsto p_{t+1}^T$ are bounded. Since the Clarke differentials at a point are compact [38], we obtain that, for all $t \in \mathbb{N}$, the sequences $T \mapsto \varphi_t^T$ and $T \mapsto \psi_t^T$ are bounded. Using Lemma 2.1, there exist an increasing function $\sigma : \mathbb{N}_* \to \mathbb{N}_*$, $\lambda_0 \in \mathbb{R}$, $p_{t+1} \in \mathbb{R}^{n*}$, $\varphi_t \in \partial_1 \phi_t(\hat{x}_t, \hat{u}_t)$ and $\psi_t \in \partial_2 \phi_t(\hat{x}_t, \hat{u}_t)$ such that the following equalities hold for all $t \in \mathbb{N}$:

$$\left.\begin{aligned} \lim_{T\to+\infty} \lambda_0^{\sigma(T)} &= \lambda_0 \\ \lim_{T\to+\infty} p_{t+1}^{\sigma(T)} &= p_{t+1} \\ \lim_{T\to+\infty} \varphi_t^{\sigma(T)} &= \varphi_t \\ \lim_{T\to+\infty} \psi_t^{\sigma(T)} &= \psi_t . \end{aligned}\right\} \tag{2.13}$$

Taking $T \to +\infty$, from $\|(\lambda_0^{\sigma(T)}, p_1^{\sigma(T)})\| = 1$ and from (2.13), we obtain $\|(\lambda_0, p_1)\| = 1$ that ensures the conclusion (i).

From (2.9) and (2.13) we obtain the conclusion (ii). From (2.10) and (2.13) we obtain the conclusion (iii). From (1.13) and (2.13) we obtain

$$p_{t+1} = p_t \circ (D_1 f_t(\hat{x}_t, \hat{u}_t))^{-1} - \lambda_0 \varphi_t \circ (D_1 f_t(\hat{x}_t, \hat{u}_t))^{-1}$$

which implies the conclusion (iv).

From (2.11) and 2.13) we obtain, for all $v_t \in T_{U_t}(\hat{u}_t)$,

$$\langle \lambda_0 \psi_t + p_{t+1} \circ D_2 f_t(\hat{x}_t, \hat{u}_t), v_t \rangle \leq 0$$

which implies $\lambda_0 \psi_t + p_{t+1} \circ D_2 f_t(\hat{x}_t, \hat{u}_t) \in N_{U_t}(\hat{u}_t)$ with $\psi_t \in \partial_2 \phi_t(\hat{x}_t, \hat{u}_t)$, which implies the conclusion (v). □

When $\hat{u}_t \in \operatorname{int} U_t$, using Corollary 1.3 instead of Proposition 1.5 and Corollary 1.4 instead of Proposition 1.6 and proceeding as in the proof of Theorem 2.3 we obtain the following result.

Theorem 2.4. *Let $(\underline{\hat{x}}, \underline{\hat{u}})$ be a solution of $(\mathscr{P}_a^n)$, or of $(\mathscr{P}_a^s)$, or of $(\mathscr{P}_a^o)$, or of $(\mathscr{P}_a^w)$ for $a \in \{e, i\}$. We assume that the following conditions are fulfilled:*

(a) For all $t \in \mathbb{N}$, ϕ_t is Lipschitzian on a neighborhood of $(\hat{x}_t, \hat{u}_t)$ and regular at $(\hat{x}_t, \hat{u}_t)$.
(b) For all $t \in \mathbb{N}$, f_t is strictly differentiable at $(\hat{x}_t, \hat{u}_t)$.
(c) For all $t \in \mathbb{N}$, $\hat{u}_t \in \operatorname{int} U_t$.

(d) For all $t \in \mathbb{N}$, the partial differential $D_1 f_t(\hat{x}_t, \hat{u}_t)$ is invertible.

Then there exist $\lambda_0 \in \mathbb{R}$, $(p_t)_{t\in\mathbb{N}_*} \in (\mathbb{R}^{n*})^{\mathbb{N}_*}$ which satisfy the following conditions:

(i) $(\lambda_0, p_1) \neq (0,0)$.
(ii) $\lambda_0 \geq 0$.
(iii) For all $t \in \mathbb{N}$, $p_{t+1} \geq 0$ and $\langle p_{t+1}, f_t(\hat{x}_t, \hat{u}_t) - \hat{x}_{t+1}\rangle = 0$ when $a = i$.
(iv) For all $t \in \mathbb{N}_*$, $p_t \in \partial_2 H_t(\hat{x}_t, \hat{u}_t, p_{t+1}, \lambda_0)$.
(v) For all $t \in \mathbb{N}$, $0 \in \partial_2 H_t(\hat{x}_t, \hat{u}_t, p_{t+1}, \lambda_0)$.

2.2.2 A Condition of Positivity

All the weak Pontryagin principles of the previous subsection use the condition of invertibility of $D_1 f_t(\hat{x}_t, \hat{u}_t)$. In this subsection, to avoid this condition of invertibility, we introduce the following positivity condition:

$$\left.\begin{array}{ll} \forall i,j \in \{1,\ldots,n\}, & \dfrac{\partial f_t^j(\hat{x}_t,\hat{u}_t)}{\partial x^i} \geq 0 \\ \forall j \in \{1,\ldots,n\}, & \dfrac{\partial f_t^j(\hat{x}_t,\hat{u}_t)}{\partial x^j} > 0. \end{array}\right\} \tag{2.14}$$

Note that this condition does not imply the invertibility of $D_1 f_t(\hat{x}_t, \hat{u}_t)$ when $n > 1$. To see that it suffices to consider the case where $\frac{\partial f_t^j(\hat{x}_t,\hat{u}_t)}{\partial x^i} = 1$ for all i, j, the condition (2.14) is fulfilled, and $D_1 f_t(\hat{x}_t, \hat{u}_t)$ is not invertible since its rank is equal to 1. In this context, this condition was introduced for the first time in the paper of Blot [11].

The following elementary lemma will be very useful.

Lemma 2.3. *Under (2.14), setting $\varrho_t := \min_{1\leq j\leq n} \frac{\partial f_t^j(\hat{x}_t,\hat{u}_t)}{\partial x^j} > 0$, the following assertions hold:*

(i) For all $y \in \mathbb{R}^n_+$, $D_1 f_t(\hat{x}_t, \hat{u}_t).y \geq \varrho_t y$.
(ii) For all $\pi \in \mathbb{R}^{n*}_+$, $\pi \circ D_1 f_t(\hat{x}_t, \hat{u}_t) \geq \varrho_t \pi$.

Proof. $(e_i)_{1\leq i\leq n}$ denotes the canonical basis of R^n, and $(e_i^*)_{1\leq i\leq n}$ its dual basis.

(i) When $y \in \mathbb{R}^n_+$, and $j \in \{1,\ldots,n\}$ we have

$$\langle e_j^*, D_1 f_t(\hat{x}_t,\hat{u}_t).y\rangle = \sum_{i=1}^{n} \frac{\partial f_t^j(\hat{x}_t,\hat{u}_t)}{\partial x^i} y^i \geq \frac{\partial f_t^j(\hat{x}_t,\hat{u}_t)}{\partial x^j} y^j + 0 \geq \varrho_t y^j,$$

that means $D_1 f_t(\hat{x}_t, \hat{u}_t).y \geq \varrho_t y$ for the natural order of $\mathbb{R}^n$.

(ii) Let $\pi \in \mathbb{R}^{n*}_+$. For all $y \in \mathbb{R}^n_+$, after (i), we have $D_1 f_t(\hat{x}_t, \hat{u}_t).y \geq \varrho_t y$, and since $\pi \geq 0$, we have

$$\pi \circ D_1 f_t(\hat{x}_t, \hat{u}_t).y = \pi(D_1 f_t(\hat{x}_t, \hat{u}_t).y) \geq \pi(\varrho_t y) = \varrho_t \pi(y),$$

that means $\pi \circ D_1 f_t(\hat{x}_t, \hat{u}_t) \geq \varrho_t \pi$ for the order of $\mathbb{R}^{n*}$. □

The following remark contains elementary facts on the orders of $\mathbb{R}^n$ and of $\mathbb{R}^{n*}$ which will be very useful.

Remark 2.1. When $\|.\|$ is one of the usual norms of $\mathbb{R}^n$, $\|x\|_\infty := \max_{1 \leq i \leq n} |x^i|$, $\|x\|_1 := \sum_{i=1}^{n} |x^i|$, $\|x\|_2 := \sqrt{\sum_{i=1}^{n} |x^i|^2}$, it satisfies the following property:

$$\forall x, y \in \mathbb{R}^n, 0 \leq x \leq y \Longrightarrow \|x\| \leq \|y\|.$$

It is easy to verify this property by using elementary calculations.

When $\mathbb{R}^n$ is endowed with the norm $\|.\|_\infty$, and $\mathbb{R}^{n*}$ is endowed with the norm $\|\varphi\|_* := \sup\{|\langle \varphi, x \rangle| : x \in \mathbb{R}^n, \|x\|_\infty \leq 1\}$, it is easy to see that $\|\varphi\|_* = \sum_{i=1}^{n} |\langle \varphi, e_i \rangle|$. The following property holds:

$$\forall \varphi, \psi \in \mathbb{R}^{n*}, 0 \leq \varphi \leq \psi \Longrightarrow \|\varphi\|_* \leq \|\psi\|_*.$$

To verify that, noting that $e_i \geq 0$ for all $i \in \{1, \ldots, n\}$, and then we have $0 \leq \varphi \leq \psi \Longrightarrow 0 \leq \varphi(e_i) \leq \psi(e_i)$, for all $i \in \{1, \ldots, n\}$, which implies

$$\|\varphi\|_* = \sum_{i=1}^{n} |\langle \varphi, e_i \rangle| = \sum_{i=1}^{n} \langle \varphi, e_i \rangle \leq \sum_{i=1}^{n} \langle \psi, e_i \rangle = \sum_{i=1}^{n} |\langle \psi, e_i \rangle| = \|\psi\|_*.$$

Now we can establish a weak Pontryagin principle.

Theorem 2.5. *Let $(\underline{\hat{x}}, \underline{\hat{u}})$ be a solution of $(\mathcal{P}^n_i)$, or of $(\mathcal{P}^s_i)$, or of $(\mathcal{P}^o_i)$, or of $(\mathcal{P}^w_i)$. We assume that the following conditions are fulfilled:*

(a) For all $t \in \mathbb{N}$, X_t is a nonempty open subset of $\mathbb{R}^n$.
(b) For all $t \in \mathbb{N}$,

$$U_t = \left(\bigcap_{\alpha=1}^{k^i} \{u \in \mathbb{R}^d : g_t^\alpha(u) \geq 0\} \right) \cap \left(\bigcap_{\beta=1}^{k^e} \{u \in \mathbb{R}^d : h_t^\beta(u) = 0\} \right)$$

where $g_t^\alpha : \mathbb{R}^d \to \mathbb{R}$ and $h_t^\beta : \mathbb{R}^d \to \mathbb{R}$ are continuous on a neighborhood of $\hat{u}_t$ and they are differentiable at $\hat{u}_t$, for all $\alpha \in \{1, \ldots, k^i\}$ and for all $\beta \in \{1, \ldots, k^e\}$, and we assume that $U_t \neq \emptyset$.

(c) For all $t \in \mathbb{N}$, the differentials $Dg_t^1(\hat{u}_t), \dots, Dg_t^{k^i}(\hat{u}_t)$, $Dh_t^1(\hat{u}_t), \dots, Dh_t^{k^e}(\hat{u}_t)$ are linearly independent.

(d) For all $t \in \mathbb{N}$, ϕ_t and f_t are continuous on a neighborhood of $(\hat{x}_t, \hat{u}_t)$, and they are differentiable at $(\hat{x}_t, \hat{u}_t)$.

(e) For all $t \in \mathbb{N}$, the positivity condition (2.14) is fulfilled.

Then there exist $\lambda_0 \in \mathbb{R}$, $(p_t)_{t\in\mathbb{N}_*} \in (\mathbb{R}^{n*})^{\mathbb{N}_*}$, $(\lambda_{1,t})_{t\in\mathbb{N}} \in \mathbb{R}^{\mathbb{N}}, \dots, (\lambda_{k^i,t})_{t\in\mathbb{N}} \in \mathbb{R}^{\mathbb{N}}$ $(\mu_{1,t})_{t\in\mathbb{N}} \in \mathbb{R}^{\mathbb{N}}, \dots, (\mu_{k^e,t})_{t\in\mathbb{N}} \in \mathbb{R}^{\mathbb{N}}$ which satisfy the following conditions:

(i) $(\lambda_0, p_1) \neq (0,0)$.

(ii) $\lambda_0 \geq 0$.

(iii) For all $t \in \mathbb{N}$, $p_{t+1} \geq 0$ and $\langle p_{t+1}, f_t(\hat{x}_t, \hat{u}_t) - \hat{x}_{t+1}\rangle = 0$.

(iv) For all $t \in \mathbb{N}$, for all $\alpha \in \{1, \dots, k^i\}$, $\lambda_{\alpha,t} \geq 0$.

(v) For all $t \in \mathbb{N}$, for all $\alpha \in \{1, \dots, k^i\}$, $\lambda_{\alpha,t} g_t^\alpha(\hat{u}_t) = 0$.

(vi) For all $t \in \mathbb{N}$, $p_t = p_{t+1} \circ D_1 f_t(\hat{x}_t, \hat{u}_t) + \lambda_0 D_1\phi_t(\hat{x}_t, \hat{u}_t)$.

(vii) For all $t \in \mathbb{N}$,

$$D_2 H_t(\hat{x}_t, \hat{u}_t, p_{t+1}, \lambda_0) + \sum_{\alpha=1}^{k^i} \lambda_{\alpha,t} Dg_t^\alpha(\hat{u}_t) + \sum_{\beta=1}^{k^e} \mu_{\beta,t} Dh_t^\beta(\hat{u}_t) = 0.$$

Proof. Using Proposition 1.2 and Proposition 1.4, we obtain, for all $T \in \mathbb{N}$, $T \geq 2$, the existence of $\lambda_0^T \in \mathbb{R}$ and, for all $t \in \{0, \dots, T-1\}$, the existence of a list of elements of $p_{t+1}^T \in \mathbb{R}^{n*}$ and the existence of two lists of real numbers $(\lambda_{\alpha,t}^T)_{1\leq\alpha\leq k^i}$, $(\mu_{\beta,t}^T)_{1\leq\beta\leq k^e}$ which satisfy the following conditions:

$$(\lambda_0^T, p_1^T, \dots, p_T^T, \lambda_{1,t}^T, \dots, \lambda_{k^i,T-1}^T, \mu_{1,t}^T, \dots, \mu_{k^e,T-1}^T) \neq (0, \dots, 0). \tag{2.15}$$

$$\lambda_0^T \geq 0. \tag{2.16}$$

$$p_{t+1}^T \geq 0. \tag{2.17}$$

$$\lambda_{\alpha,t}^T \geq 0, \lambda_{\alpha,t}^T g_t^\alpha(\hat{u}_t) = 0. \tag{2.18}$$

$$p_t^T = p_{t+1}^T \circ D_1 f_t(\hat{x}_t, \hat{u}_t) + \lambda_0^T D_1\phi_t(\hat{x}_t, \hat{u}_t). \tag{2.19}$$

$$\left.\begin{aligned} & p_{t+1}^T \circ D_2 f_t(\hat{x}_t, \hat{u}_t) + \lambda_0^T D_2\phi_t(\hat{x}_t, \hat{u}_t) \\ & + \sum_{\alpha=1}^{k^i} \lambda_{\alpha,t}^T Dg_t^\alpha(\hat{u}_t) + \sum_{\beta=1}^{k^e} \mu_{\beta,t}^T Dh_t^\beta(\hat{u}_t) = 0. \end{aligned}\right\} \tag{2.20}$$

From (2.19) we obtain $p_{t+1}^T \circ (\hat{x}_t, \hat{u}_t) = p_t^T - \lambda_0^T D_1\phi_t(\hat{x}_t, \hat{u}_t)$, and after Lemma 2.3 we have $p_{t+1}^T \circ D_1 f_t(\hat{x}_t, \hat{u}_t) \geq \varrho_t p_{t+1}^T$. And so we have

$$0 \leq \varrho_t p_{t+1}^T \leq p_t^T - \lambda_0^T D_1\phi_t(\hat{x}_t, \hat{u}_t),$$

which implies (cf. Remark 2.1)

$$\varrho_t \|p_{t+1}^T\|_* = \|\varrho_t p_{t+1}^T\|_* \leq \|p_t^T - \lambda_0^T D_1\phi_t(\hat{x}_t, \hat{u}_t)\|_* \leq \|p_t^T\|_* + \lambda_0^T \|D_1\phi_t(\hat{x}_t, \hat{u}_t)\|_*.$$

And so we obtain, for all $t \in \{0, \dots, T-1\}$,

$$\|p^T_{t+1}\|_* \leq \frac{1}{\varrho_t}\|p^T_t\|_* + \lambda^T_0 \frac{1}{\varrho_t}\|D_1\phi_t(\hat{x}_t, \hat{u}_t)\|_*. \tag{2.21}$$

We set

$$a_t := \frac{1}{\prod\limits_{s=1}^{t} \varrho_s} \in (0, +\infty)$$

and

$$b_t := \sum_{s=1}^{t} \frac{1}{\prod\limits_{k=s}^{t} \varrho_k} \|D_1\phi_s(\hat{x}_t, \hat{u}_s)\|_* \in (0, +\infty)$$

and we proceed by induction to obtain from (2.21) the following assertion:

$$\left.\begin{array}{l} \forall t \in \mathbb{N}, \exists a_t \in (0, +\infty), \exists b_t \in (0, +\infty), \forall T > t, \\ \|p^T_{t+1}\|_* \leq a_t \|p^T_1\|_* + b_t \lambda^T_0. \end{array}\right\} \tag{2.22}$$

From this last assertion we obtain that $(\lambda^T_0, p^T_1) = (0, 0)$ implies $(\lambda^T_0, p^T_1, \dots, p^T_T) = (0, 0, \dots, 0)$, and using (2.20) we obtain

$$\sum_{\alpha=1}^{k^i} \lambda^T_{\alpha,t} Dg^\alpha_t(\hat{u}_t) + \sum_{\beta=1}^{k^e} \mu^T_{\beta,t} Dh^\beta_t(\hat{u}_t) = 0,$$

and using assumption (c), this last equality implies that $\lambda^T_{\alpha,t} = 0$ and $\mu^T_{\beta,t} = 0$ for all $\alpha \in \{1, \dots, k^i\}$, for all $\beta \in \{1, \dots, k^e\}$ and for all $t \in \{0, \dots, T-1\}$. And so we have proven that $(\lambda^T_0, p^T_1) = (0, 0)$ implies the negation of (2.15). Then using (2.15) and the contraposition we have proven that $(\lambda^T_0, p^T_1) \neq (0, 0)$. Using the property of cone of the set of all lists of multipliers, we can normalize and obtain, for all $T \in \mathbb{N}$, $T \geq 2$:

$$\|(\lambda^T_0, p^T_1)\| = 1. \tag{2.23}$$

Then, from (2.22) and (2.23) we obtain

$$\forall t \in \mathbb{N}, \exists a_t \in (0, +\infty), \exists b_t \in (0, +\infty), \forall T > t, \|p^T_{t+1}\|_* \leq a_t + b_t$$

which implies

$$\forall t \in \mathbb{N}, \quad \sup_{T>t} \| p_{t+1}^T \|_* \leq +\infty. \tag{2.24}$$

From (2.20) we obtain

$$\begin{cases} \sum_{\alpha=1}^{k^i} \lambda_{\alpha,t}^T Dg_t^\alpha(\hat{u}_t) + \sum_{\beta=1}^{k^e} \mu_{\beta,t}^T Dh_t^\beta(\hat{u}_t) \\ = -p_{t+1}^T \circ D_2 f_t(\hat{x}_t, \hat{u}_t) - \lambda_0^T D_2\phi_t(\hat{x}_t, \hat{u}_t), \end{cases}$$

and using (2.23) and (2.24) we deduce that the sequence $T \mapsto \sum_{\alpha=1}^{k^i} \lambda_{\alpha,t}^T Dg_t^\alpha(\hat{u}_t) + \sum_{\beta=1}^{k^e} \mu_{\beta,t}^T Dh_t^\beta(\hat{u}_t)$ is bounded for all $t \in \mathbb{N}$, and then using assumption (c) and Lemma 2.2, we obtain

$$\forall t \in \mathbb{N}, \forall \alpha \in \{1, \ldots, k^i\}, \forall \beta \in \{1, \ldots, k^e\}, \sup_{T>t} |\lambda_{\alpha,t}^T| < +\infty, \sup_{T>t} |\mu_{\beta,t}^T| < +\infty. \tag{2.25}$$

And then we can conclude as in the proof of Theorem 2.1. □

When $\hat{u}_t \in \operatorname{int} U_t$, using Corollary 1.2 instead of Proposition 1.4, we obtain the following result.

Theorem 2.6. *Let $(\underline{\hat{x}}, \underline{\hat{u}})$ be a solution of $(\mathcal{P}_i^n)$, or of $(\mathcal{P}_i^s)$, or of $(\mathcal{P}_i^o)$, or of $(\mathcal{P}_i^w)$. We assume that the following conditions are fulfilled:*

(a) For all $t \in \mathbb{N}$, X_t is a nonempty open subset of $\mathbb{R}^n$.
(b) For all $t \in \mathbb{N}$, $\hat{u}_t \in \operatorname{int} U_t$.
(c) For all $t \in \mathbb{N}$, ϕ_t and f_t are continuous on a neighborhood of $(\hat{x}_t, \hat{u}_t)$ and they are differentiable at $(\hat{x}_t, \hat{u}_t)$.
(d) For all $t \in \mathbb{N}$, the condition (2.14) holds.

Then there exist $\lambda_0 \in \mathbb{R}$, $(p_t)_{t \in \mathbb{N}_*} \in (\mathbb{R}^{n*})^{\mathbb{N}_*}$ which satisfy the following conditions:

(i) $(\lambda_0, p_1) \neq (0, 0)$.
(ii) $\lambda_0 \geq 0$.
(iii) For all $t \in \mathbb{N}$, $p_{t+1} \geq 0$ and $\langle p_{t+1}, f_t(\hat{x}_t, \hat{u}_t) - \hat{x}_{t+1} \rangle = 0$.
(iv) For all $t \in \mathbb{N}$, $p_t = p_{t+1} \circ D_1 f_t(\hat{x}_t, \hat{u}_t) + \lambda_0 D_1\phi_t(\hat{x}_t, \hat{u}_t)$.
(v) For all $t \in \mathbb{N}$, $\quad D_2 H_t(\hat{x}_t, \hat{u}_t, p_{t+1}, \lambda_0) = 0$.

After the use of the multiplier rule of Halkin, we use the multiplier rule of Clarke.

Theorem 2.7. *Let $(\underline{\hat{x}}, \underline{\hat{u}})$ be a solution of $(\mathcal{P}_i^n)$, or of $(\mathcal{P}_i^s)$, or of $(\mathcal{P}_i^o)$, or of $(\mathcal{P}_i^w)$. We assume that the following conditions are fulfilled:*

(a) *For all* $t \in \mathbb{N}$, ϕ_t *is Lipschitzian on a neighborhood of* $(\hat{x}_t, \hat{u}_t)$ *and regular at* $(\hat{x}_t, \hat{u}_t)$.
(b) *For all* $t \in \mathbb{N}$, f_t *is strictly differentiable at* $(\hat{x}_t, \hat{u}_t)$.
(c) *For all* $t \in \mathbb{N}$, U_t *is closed and Clarke-regular at* $\hat{u}_t$.
(d) *For all* $t \in \mathbb{N}$, *the positivity condition (2.14) holds.*

Then there exist $\lambda_0 \in \mathbb{R}$, $(p_t)_{t\in\mathbb{N}_*} \in (\mathbb{R}^{n*})^{\mathbb{N}_*}$ *which satisfy the following conditions:*

(i) $(\lambda_0, p_1) \neq (0, 0)$.
(ii) $\lambda_0 \geq 0$.
(iii) *For all* $t \in \mathbb{N}$, $p_{t+1} \geq 0$ *and* $\langle p_{t+1}, f_t(\hat{x}_t, \hat{u}_t - \hat{x}_{t+1}\rangle = 0$.
(iv) *For all* $t \in \mathbb{N}_*$, $p_t \in \partial_2 H_t(\hat{x}_t, \hat{u}_t, p_{t+1}, \lambda_0)$.
(v) *For all* $t \in \mathbb{N}$, $\partial_2 H_t(\hat{x}_t, \hat{u}_t, p_{t+1}, \lambda_0) \cap N_{U_t}(\hat{u}_t) \neq \emptyset$, *where* $N_{U_t}(\hat{u}_t)$ *is the normal cone of* U_t *at* $\hat{u}_t$.

Proof. Using Proposition 1.2 and Proposition 1.6 we obtain the assertions (2.15), (2.16), and (2.17) inside the proof of Theorem 2.5 and the assertions (2.10) and (2.11) inside the proof of Theorem 2.3.

Since the Clarke differentials $\partial_1\phi_t(\hat{x}_t, \hat{u}_t)$ and $\partial_2\phi_t(\hat{x}_t, \hat{u}_t)$ are compact sets, they are bounded sets, and so we have

$$c_t : = \sup\{\|\varphi\|_* : \varphi \in \partial_1\phi_t(\hat{x}_t, \hat{u}_t)\} < +\infty$$
$$d_t : = \sup\{\|\psi\|_* : \psi \in \partial_2\phi_t(\hat{x}_t, \hat{u}_t)\} < +\infty.$$

Then, from (2.10), we obtain $p_{t+1}^T \circ D_1 f_t(\hat{x}_t, \hat{u}_t) = p_t^T - \lambda_0^T \varphi_t^T$ which implies $\varrho_t p_{t+1}^T \leq p_t^T - \lambda_0^T \varphi_t^T$, where ϱ_t is defined in the proof of Theorem 2.5, which implies (cf. Remark 2.1) $\varrho_t \|p_{t+1}^T\|_* \leq \|p_t^T\|_* + \lambda_0^T c_t$, i.e.,

$$\|p_{t+1}^T\|_* \leq \frac{1}{\varrho_t}\|p_t^T\|_* + \frac{c_t}{\varrho_t}\lambda_0^T.$$

And after that, we can proceed as in the proof of Theorem 2.5. □

When $\hat{u}_t \in \operatorname{int} U_t$, using Corollary 1.4 instead of Proposition 1.6 and proceeding as in the proof of Theorem 2.7 we obtain the following result.

Theorem 2.8. *Let* $(\underline{\hat{x}}, \underline{\hat{u}})$ *be a solution of* $(\mathscr{P}_i^n)$, *or of* $(\mathscr{P}_i^s)$, *or of* $(\mathscr{P}_i^o)$, *or of* $(\mathscr{P}_i^w)$. *We assume that the following conditions are fulfilled:*

(a) *For all* $t \in \mathbb{N}$, ϕ_t *is Lipschitzian on a neighborhood of* $(\hat{x}_t, \hat{u}_t)$ *and regular at* $(\hat{x}_t, \hat{u}_t)$.
(b) *For all* $t \in \mathbb{N}$, f_t *is strictly differentiable at* $(\hat{x}_t, \hat{u}_t)$.
(c) *For all* $t \in \mathbb{N}$, $\hat{u}_t \in \operatorname{int} U_t$.
(d) *For all* $t \in \mathbb{N}$, *the positivity condition (2.14) holds.*

Then there exist $\lambda_0 \in \mathbb{R}$, $(p_t)_{t\in\mathbb{N}_*} \in (\mathbb{R}^{n*})^{\mathbb{N}_*}$ *which satisfy the following conditions:*

(*i*) $(\lambda_0, p_1) \neq (0,0)$.
(*ii*) $\lambda_0 \geq 0$.
(*iii*) *For all* $t \in \mathbb{N}$, $p_{t+1} \geq 0$ *and* $\langle p_{t+1}, f_t(\hat{x}_t, \hat{u}_t) - \hat{x}_{t+1} \rangle = 0$.
(*iv*) *For all* $t \in \mathbb{N}_*$, $p_t \in \partial_2 H_t(\hat{x}_t, \hat{u}_t, p_{t+1}, \lambda_0)$.
(*v*) *For all* $t \in \mathbb{N}$, $0 \in \partial_2 H_t(\hat{x}_t, \hat{u}_t, p_{t+1}, \lambda_0)$.

2.3 Strong Pontryagin Principles in Infinite Horizon

In this section, to establish strong Pontryagin principles in infinite horizon, in a first subsection, we use the invertibility condition, in a second subsection we use the positivity condition, and in a third subsection, we use a new condition that we call a condition of partial submersion.

2.3.1 The Invertibility Condition

In a first time we use a consequence of a result of Michel, Proposition 1.7 and Proposition 1.8.

Theorem 2.9. *Let* $(\underline{\hat{x}}, \underline{\hat{u}})$ *be a solution of* $(\mathcal{P}_a^n)$, *or of* $(\mathcal{P}_a^s)$, *or of* $(\mathcal{P}_a^o)$, *or of* $(\mathcal{P}_a^w)$ *when* $a \in \{e, i\}$. *We assume that the following conditions are fulfilled:*

(*a*) *For all* $t \in \mathbb{N}$, X_t *is a nonempty open convex subset of* $\mathbb{R}^n$ *and* U_t *is a nonempty subset of* $\mathbb{R}^d$.
(*b*) *For all* $t \in \mathbb{N}$, *the functions* ϕ_t *and* f_t *are differentiable with respect to the first vector variable.*
(*c*) *For all* $t \in \mathbb{N}$, *for all* $(x_t, x_{t+1}) \in X_t \times X_{t+1}$, $coA_t'(x_t, x_{t+1}) \subset B_t'(x_t, x_{t+1})$.
(*d*) *For all* $t \in \mathbb{N}$, *the partial differential* $D_1 f_t(\hat{x}_t, \hat{u}_t)$ *is invertible.*

Then there exist $\lambda_0^T \in \mathbb{R}$ *and* $(p_{t+1})_{t\in\mathbb{N}} \in (\mathbb{R}^{n*})^{\mathbb{N}}$ *which satisfy the following conditions:*

(*i*) $(\lambda_0^T, p_1) \neq (0,0)$.
(*ii*) $\lambda_0^T \geq 0$.
(*iii*) *For all* $t \in \mathbb{N}_*$, $p_t^T = D_1 H_t(\hat{x}_t, \hat{u}_t, p_{t+1}, \lambda_0)$.
(*iv*) *For all* $t \in \mathbb{N}$, $H_t(\hat{x}_t, \hat{u}_t, p_{t+1}, \lambda_0^T) = \max_{u \in U_t} H_t(\hat{x}_t, u, p_{t+1}, \lambda_0^T)$.

Proof. **The case** $a = e$. From Proposition 1.2 we can use Proposition 1.7 which provides, for all $T \in \mathbb{N}$, $T \geq 2$, a real number λ_0^T and elements of the dual space of $\mathbb{R}^n$, $p_1^T, \ldots, p_T^T$, which satisfy the following conditions:

$$(\lambda_0^T, p_1^T, \ldots, p_T^T) \neq (0, 0, \ldots, 0). \tag{2.26}$$

$$\lambda_0^T \geq 0. \tag{2.27}$$

$$\forall t \in \{1, \ldots, T-1\}, p_t^T = p_{t+1}^T \circ D_1 f_t(\hat{x}_t, \hat{u}_t) + \lambda_0^T D_1 \phi_t(\hat{x}_t, \hat{u}_t). \tag{2.28}$$

$$\forall t \in \{1, \ldots, T-1\}, \forall u \in U_t, H_t\left(\hat{x}_t, \hat{u}_t, p_{t+1}^T, \lambda_0^T\right) \geq H_t\left(\hat{x}_t, u, p_{t+1}^T, \lambda_0^T\right). \tag{2.29}$$

Proceeding as in the proof of Theorem 2.1, we obtain the relations (2.3) and (2.4) which say that the sequences $T \mapsto \lambda_0^T$ and $T \mapsto p_{t+1}^T$ (for all $t \in \mathbb{N}$) are bounded with the additional condition $\|(\lambda_0^T, p_1^T)\| = 1$. And so we can use Lemma 2.1 and we can assert that there exist an increasing function $\sigma : \mathbb{N}_* \to \mathbb{N}_*$, $\lambda_0 \in \mathbb{R}$, $p_{t+1} \in \mathbb{R}^{n*}$ for all $t \in \mathbb{N}$, such that the following relations hold:

$$\lim_{T \to +\infty} \lambda_0^{\sigma(T)} = \lambda_0, \quad \lim_{T \to +\infty} p_{t+1}^{\sigma(T)} = p_{t+1}$$

for all $t \in \mathbb{N}$. Using the continuity of the norm we obtain $\|(\lambda_0, p_1)\| = 1$. Using the continuity of the functions inside the relations (2.27)–(2.29), we obtain the conclusions of the theorem.

The case $a = i$. Our strategy is to use the first case. For all $t \in \mathbb{N}$ we introduce the function $\hat{f}_t : X_t \times U_t \to X_{t+1}$ by setting

$$\hat{f}_t(x_t, u_t) := f_t(x_t, u_t) + (\hat{x}_{t+1} - f_t(\hat{x}_t, \hat{u}_t)). \tag{2.30}$$

We denote by $\mathrm{Adm}_\eta^e(\hat{f})$ the set of all processes $(\underline{x}, \underline{u}) \in \prod_{t \in \mathbb{N}} X_t \times \prod_{t \in \mathbb{N}}$ such that $x_{t+1} = \hat{f}_t(x_t, u_t)$ for all $t \in \mathbb{N}$, and we denote by Adm_η^i the set of all processes $(\underline{x}, \underline{u}) \in \prod_{t \in \mathbb{N}} X_t \times \prod_{t \in \mathbb{N}}$ such that $x_{t+1} \leq f_t(x_t, u_t)$ for all $t \in \mathbb{N}$. Since $\hat{x}_{t+1} \leq f_t(\hat{x}_t, \hat{u}_t)$, we have, for all $(\underline{x}, \underline{u}) \in \mathrm{Adm}_\eta^e(\hat{f})$, $x_{t+1} = \hat{f}_t(x_t, u_t) \leq f_t(x_t, u_t)$, which implies

$$\mathrm{Adm}_\eta^e(\hat{f}) \subset \mathrm{Adm}_\eta^i. \tag{2.31}$$

We denote by $\mathrm{Dom}_\eta^e(J, \hat{f})$ the set of all $(\underline{x}, \underline{u}) \in \mathrm{Adm}_\eta^e(\hat{f})$ which belong to $\mathrm{Dom}_\eta^e(J)$ (cf. Sect. 1.2). Using (2.34), it is clear that we have

$$\mathrm{Dom}_\eta^e(J, \hat{f}) \subset \mathrm{Dom}_\eta^i(J). \tag{2.32}$$

Note that $(\underline{\hat{x}}, \underline{\hat{u}}) \in \mathrm{Adm}_\eta^e(\hat{f})$, and consequently that $(\underline{\hat{x}}, \underline{\hat{u}}) \in \mathrm{Dom}_\eta^e(J, \hat{f})$ when $(\underline{\hat{x}}, \underline{\hat{u}}) \in \mathrm{Dom}_\eta^i(J)$.

We fix $k \in \{n, s, o, w\}$ and we denote by $(\mathscr{P}_e^k(\hat{f}))$ the problem $(\mathscr{P}_e^k)$ where we have replaced (DE) by $x_{t+1} = \hat{f}_t(x_t, u_t)$. Note that the criterion of this problem is the same as the criterion of $(\mathscr{P}_i^k)$. And so, using the previous inclusions, we see that if $(\underline{\hat{x}}, \underline{\hat{u}})$ is a solution of $(\mathscr{P}_i^k)$ then it is also a solution of $(\mathscr{P}_e^k(\hat{f}))$. We see that the assumptions on f_t imply the same assumptions on $\hat{f}_t$, and so we can apply the first case to $(\mathcal{P}_e^k(\hat{f}))$. After that, it suffices to translate the conclusions on $(\mathcal{P}_e^k(\hat{f}))$ into conclusions on $(\mathcal{P}_i^k)$. If we denote by $\hat{H}_t$ the Hamiltonian of $(\mathcal{P}_e^k(\hat{f}))$, H_t being the Hamiltonian of $(\mathcal{P}_i^k)$, we see that the difference $\hat{H}_t - H_t$ is a constant which is independent of x_t and u_t, which implies that the adjoint equation of $(\mathcal{P}_e^k(\hat{f}))$ is exactly the adjoint equation of $(\mathcal{P}_i^k)$, and the strong maximum principle of $(\mathcal{P}_e^k(\hat{f}))$ implies this one of $(\mathcal{P}_i^k)$. □

This theorem was established in [16]. There exist other versions in [13]. In this last paper an analogous version for systems governed by (DI) is stated. But the proof given for the case of (DI) is not very explicative. And so, we provide an original proof of the theorem in the case of (DI).

Theorem 2.10. *Let $(\underline{\hat{x}}, \underline{\hat{u}})$ be a solution of $(\mathcal{P}_a^n)$, or of $(\mathcal{P}_a^s)$, or of $(\mathcal{P}_a^o)$, or of $(\mathcal{P}_a^w)$, when $a \in \{e, i\}$. We assume that the following conditions are fulfilled:*

(a) For all $t \in \mathbb{N}$, X_t is a nonempty open subset of $\mathbb{R}^n$, and U_t is a nonempty subset of $\mathbb{R}^d$.
(b) For all $t \in \mathbb{N}$, $\phi_t \in C^0(X_t \times U_t, \mathbb{R})$ and, for all $(x,u) \in \mathbb{R}^n \times U_t$, the partial differential $D_1\phi_t(x,u)$ exists and $D_1\phi_t \in C^0(X_t \times U_t, \mathbb{R}^{n*})$.
(c) For all $t \in \mathbb{N}$, $f_t \in C^0(X_t \times U_t, \mathbb{R}^n)$ and, for all $(x,u) \in \mathbb{R}^n \times U_t$, the partial differential $D_1 f_t(x,u)$ exists and $D_1\phi_t \in C^0(X_t \times U_t, \mathcal{L}(\mathbb{R}^n, \mathbb{R}^n))$.
(d) For all $t \in \mathbb{N}$, for all $x \in X_t$, for all $u, v \in U_t$, for all $r \in [0,1]$, there exists $w \in U_t$ such that

$$\begin{cases} \phi_t(x,w) \geq (1-r)\phi_t(x,u) + r\phi_t(x,v) \\ f_t(x,w) = (1-r)f_t(x,u) + rf_t(x,v). \end{cases}$$

Then there exist $\lambda_0 \in \mathbb{R}$, $(p_{t+1})_{t\in\mathbb{N}} \in (\mathbb{R}^{n*})^N$ which satisfy the following conditions:

(i) λ_0 and $(p_{t+1})_{t\in\mathbb{N}}$ are not simultaneously equal to zero.
(ii) $p_t = D_1 H_t(\hat{x}_t, \hat{u}_t, p_{t+1}, \lambda_0)$ for all $t \in \mathbb{N}$.
(iii) $H_t(\hat{x}_t, \hat{u}_t, p_{t+1}, \lambda_0) = \max_{u\in U_t} H_t(\hat{x}_t, u, p_{t+1}, \lambda_0)$ for all $t \in \mathbb{N}$.

Proof. The proof is similar to this one of Theorem 2.9 replacing the use of Proposition 1.7 by the use of Proposition 1.9 when $a = e$ and the use of Proposition 1.8 by the use of Proposition 1.10 when $a = i$. □

2.3.2 A Condition of Positivity

In this subsection, we use the positivity condition already used in Sect. 2.2 to obtain weak Pontryagin principles, in order to obtain strong Pontryagin principles.

Theorem 2.11. *Let $(\underline{\hat{x}}, \underline{\hat{u}})$ be a solution of $(\mathscr{P}_i^n)$, or of $(\mathscr{P}_i^s)$, or of $(\mathscr{P}_i^o)$, or of $(\mathscr{P}_i^w)$. We assume that the following conditions are fulfilled:*

(a) For all $t \in \mathbb{N}$, X_t is nonempty and convex, $\hat{x}_t \in \operatorname{int} X_t$, and U_t is nonempty.
(b) For all $t \in \mathbb{N}$, the partial functions $\phi_t(., \hat{u}_t)$ and $f_t(., \hat{u}_t)$ are continuous on a neighborhood of $\hat{x}_t$ and differentiable at $\hat{x}_t$.
(c) For all $t \in \mathbb{N}$, for all $(x_t, x_{t+1}) \in X_t \times X_{t+1}$, $co A_t(x_t, x_{t+1}) \subset B_t(x_t, x_{t+1})$.
(d) For all $t \in \mathbb{N}$, for all $i, j \in \{1, \ldots, n\}$, $\frac{\partial f^i(\hat{x}_t, \hat{u}_t)}{\partial x_t^j} \geq 0$, and for all $j \in \{1, \ldots, n\}$, $\frac{\partial f^i(\hat{x}_t, \hat{u}_t)}{\partial x_t^j} > 0$.

Then there exist $\lambda_0 \in \mathbb{R}$, $(p_{t+1})_{t \in \mathbb{N}} \in (\mathbb{R}^{n*})^N$ which satisfy the following conditions:

(i) $(\lambda_0, p_1) \neq (0, 0)$.
(ii) $\lambda_0 \geq 0$.
(iii) For all $t \in \mathbb{N}$, $p_{t+1} \geq 0$ and $\langle p_{t+1}, f_t(\hat{x}_t, \hat{u}_t) - \hat{x}_{t+1} \rangle = 0$.
(iv) For all $t \in \mathbb{N}_*$, $p_t = D_1 H_t(\hat{x}_t, \hat{u}_t, p_{t+1}, \lambda_0)$.
(v) For all $t \in \mathbb{N}$, $H_t(\hat{x}_t, \hat{u}_t, p_{t+1}, \lambda_0^T) = \max_{u \in U_t} H_t(\hat{x}_t, u, p_{t+1}, \lambda_0)$.

Proof. Using Propositions 1.2 and 1.8, we obtain the existence, for all $T \in \mathbb{N}$, $T \geq 2$, of $\lambda_0^T \in \mathbb{R}$ and of $p_1^T, \ldots, p_T^T \in \mathbb{R}^{n*}$ which satisfy the conclusions of Proposition 1.8. Then using Lemma 2.3 and reasoning as in the proof of Theorem 2.5, we obtain the relations (2.22), (2.23), and (2.24). Then using Lemma 2.1, we obtain the existence of a strictly increasing function $\sigma : \mathbb{N}_* \to \mathbb{N}_*$, of $\lambda_0 \in \mathbb{R}$ and of a sequence $(p_{t+1})_{t \in \mathbb{N}}$ in $\mathbb{R}^{n*}$ such that $\lim_{T \to +\infty} \lambda_0^{\sigma(T)} = \lambda_0$, and $\lim_{T \to +\infty} p_{t+1}^{\sigma(T)} = p_{t+1}$ for all $t \in \mathbb{N}$. And then, from the conclusions of Proposition 1.8, we obtain the conclusion of this theorem by taking $T \to +\infty$. □

In the previous theorem, we have only considered problems which are governed by (DI). In the following theorems, we consider problems governed by (DE).

Theorem 2.12. *Let $(\underline{\hat{x}}, \underline{\hat{u}})$ be a solution of $(\mathscr{P}_e^o)$. We assume that the following conditions are fulfilled:*

(a) For all $t \in \mathbb{N}$, X_t is nonempty and convex, $\hat{x}_t \in \operatorname{int} X_t$, and U_t is nonempty.
(b) For all $t \in \mathbb{N}$, the partial functions $\phi_t(., \hat{u}_t)$ and $f_t(., \hat{u}_t)$ are continuous on a neighborhood of $\hat{x}_t$ and differentiable at $\hat{x}_t$.
(c) For all $t \in \mathbb{N}$, for all $(x_t, x_{t+1}) \in X_t \times X_{t+1}$, $co A_t(x_t, x_{t+1}) \subset B_t(x_t, x_{t+1})$.
(d) For all $t \in \mathbb{N}$, for all $i, j \in \{1, \ldots, n\}$, $\frac{\partial f^i(\hat{x}_t, \hat{u}_t)}{\partial x_t^j} \geq 0$, and for all $j \in \{1, \ldots, n\}$, $\frac{\partial f^i(\hat{x}_t, \hat{u}_t)}{\partial x_t^j} > 0$.
(e) For all $t \in \mathbb{N}$, for all $u_t \in U_t$, the partial function $\phi_t(., u_t)$ is increasing.
(f) For all $t \in \mathbb{N}$, for all $u_t \in U_t$, the partial function $f_t(., u_t)$ is increasing.

Then there exist $\lambda_0 \in \mathbb{R}$, $(p_{t+1})_{t \in \mathbb{N}} \in (\mathbb{R}^{n*})^N$ which satisfy the following conditions:

(i) $(\lambda_0, p_1) \neq (0, 0)$.
(ii) $\lambda_0 \geq 0$.
(iii) For all $t \in \mathbb{N}$, $p_{t+1} \geq 0$ and $\langle p_{t+1}, f_t(\hat{x}_t, \hat{u}_t) - \hat{x}_{t+1}\rangle = 0$.
(iv) For all $t \in \mathbb{N}_*$, $p_t = D_1 H_t(\hat{x}_t, \hat{u}_t, p_{t+1}, \lambda_0)$.
(v) For all $t \in \mathbb{N}$, $H_t(\hat{x}_t, \hat{u}_t, p_{t+1}, \lambda_0^T) = \max_{u \in U_t} H_t(\hat{x}_t, u, p_{t+1}, \lambda_0)$.

Proof. Noting that (CA1) = (e) and that (CA2) = (f), we can use Theorem 1.1 and we can assert that $(\underline{\hat{x}}, \underline{\hat{u}})$ is a solution of $(\mathscr{P}_i^o)$. And then we conclude by using Theorem 2.11. □

Theorem 2.13. *Let $(\underline{\hat{x}}, \underline{\hat{u}})$ be a solution of $(\mathscr{P}_e^o)$. We assume that the following conditions are fulfilled:*

(a) *For all $t \in \mathbb{N}$, X_t is nonempty and convex, $\hat{x}_t \in \operatorname{int} X_t$, and U_t is nonempty.*
(b) *For all $t \in \mathbb{N}$, the partial functions $\phi_t(., \hat{u}_t)$ and $f_t(., \hat{u}_t)$ are continuous on a neighborhood of $\hat{x}_t$ and differentiable at $\hat{x}_t$.*
(c) *For all $t \in \mathbb{N}$, for all $(x_t, x_{t+1}) \in X_t \times X_{t+1}$, $coA_t(x_t, x_{t+1}) \subset B_t(x_t, x_{t+1})$.*
(d) *For all $t \in \mathbb{N}$, for all $i, j \in \{1, \ldots, n\}$, $\frac{\partial f^i(\hat{x}_t, \hat{u}_t)}{\partial x_t^j} \geq 0$, and for all $j \in \{1, \ldots, n\}$, $\frac{\partial f^i(\hat{x}_t, \hat{u}_t)}{\partial x_t^j} > 0$.*
(e) *For all $t \in \mathbb{N}$, for all $x_t \in X_t$, the partial function $\phi_t(x_t, .)$ is increasing.*
(f) *For all $t \in \mathbb{N}$, for all $(y_{t+1}, y_t, u_t \in X_{t+1} \times X_t \times U_t$ such that $y_{t+1} \leq f_t(y_t, u_t)$, there exists $v_t \in U_t$ such that $v_t \geq u_t$ and $y_{t+1} = f_t(y_t, v_t)$.*

Then there exist $\lambda_0 \in \mathbb{R}$, $(p_{t+1})_{t \in \mathbb{N}} \in (\mathbb{R}^{n*})^N$ which satisfy the following conditions:

(i) $(\lambda_0, p_1) \neq (0, 0)$.
(ii) $\lambda_0 \geq 0$.
(iii) For all $t \in \mathbb{N}$, $p_{t+1} \geq 0$ and $\langle p_{t+1}, f_t(\hat{x}_t, \hat{u}_t) - \hat{x}_{t+1}\rangle = 0$.
(iv) For all $t \in \mathbb{N}_*$, $p_t = D_1 H_t(\hat{x}_t, \hat{u}_t, p_{t+1}, \lambda_0)$.
(v) For all $t \in \mathbb{N}$, $H_t(\hat{x}_t, \hat{u}_t, p_{t+1}, \lambda_0^T) = \max_{u \in U_t} H_t(\hat{x}_t, u, p_{t+1}, \lambda_0)$.

Proof. Noting that (CA4) = (f) and that (CA5) = (e), we can use Theorem 1.2 and we can assert that $(\underline{\hat{x}}, \underline{\hat{u}})$ is a solution of $(\mathscr{P}_i^o)$. And then we conclude by using Theorem 2.11. □

Theorem 2.14. *Let $(\underline{\hat{x}}, \underline{\hat{u}})$ be a solution of $(\mathscr{P}_e^n)$. We assume that the following conditions are fulfilled:*

(a) *For all $t \in \mathbb{N}$, X_t is nonempty and convex, $\hat{x}_t \in \operatorname{int} X_t$, and U_t is nonempty.*
(b) *For all $t \in \mathbb{N}$, the partial functions $\phi_t(., \hat{u}_t)$ and $f_t(., \hat{u}_t)$ are continuous on a neighborhood of $\hat{x}_t$ and differentiable at $\hat{x}_t$.*
(c) *For all $t \in \mathbb{N}$, for all $(x_t, x_{t+1}) \in X_t \times X_{t+1}$, $coA_t(x_t, x_{t+1}) \subset B_t(x_t, x_{t+1})$.*
(d) *For all $t \in \mathbb{N}$, for all $i, j \in \{1, \ldots, n\}$, $\frac{\partial f^i(\hat{x}_t, \hat{u}_t)}{\partial x_t^j} \geq 0$, and for all $j \in \{1, \ldots, n\}$, $\frac{\partial f^i(\hat{x}_t, \hat{u}_t)}{\partial x_t^j} > 0$.*
(e) *For all $t \in \mathbb{N}$, for all $u_t \in U_t$, the partial function $\phi_t(., u_t)$ is increasing.*

(f) *For all* $t \in \mathbb{N}$, *for all* $u_t \in U_t$, *the partial function* $f_t(., u_t)$ *is increasing.*
(g) *For all* $t \in \mathbb{N}$, $\phi_t \geq 0$.
(h) *For all* $t \in \mathbb{N}$, *for all* $z_t \in X_t$, *there exists* $s \in \mathbb{N}_*$ *and there exists* $(v_t, \dots, v_{t+s-1}) \in \prod_{j=0}^{s-1} U_{t+j}$ *such that by setting* $z_{t+j+1} := f_{t+j}(z_{t+j}, v_{t+j})$ *for* $j \in \{0, \dots, s-1\}$ *we have* $z_{t+s} = \hat{x}_{t+s}$.

Then there exist $\lambda_0 \in \mathbb{R}$, $(p_{t+1})_{t \in \mathbb{N}} \in (\mathbb{R}^{n*})^N$ *which satisfy the following conditions:*

(i) $(\lambda_0, p_1) \neq (0, 0)$.
(ii) $\lambda_0 \geq 0$.
(iii) *For all* $t \in \mathbb{N}$, $p_{t+1} \geq 0$ *and* $\langle p_{t+1}, f_t(\hat{x}_t, \hat{u}_t) - \hat{x}_{t+1} \rangle = 0$.
(iv) *For all* $t \in \mathbb{N}_*$, $p_t = D_1 H_t(\hat{x}_t, \hat{u}_t, p_{t+1}, \lambda_0)$.
(v) *For all* $t \in \mathbb{N}$, $H_t(\hat{x}_t, \hat{u}_t, p_{t+1}, \lambda_0^T) = \max_{u \in U_t} H_t(\hat{x}_t, u, p_{t+1}, \lambda_0)$.

Proof. Note that (CA 1) = (e), (CA 2) = (f), (CA 3) = (g), and (CA, $(\underline{\hat{x}}, \underline{\hat{u}})$) = (h). And then we can use Theorem 1.3 to assert that $(\underline{\hat{x}}, \underline{\hat{u}})$ is also a solution of $(\mathcal{P}_i^n)$. We conclude by using Theorem 2.11. □

Theorem 2.15. *Let* $(\underline{\hat{x}}, \underline{\hat{u}})$ *be a solution of* $(\mathcal{P}_e^n)$. *We assume that the following conditions are fulfilled:*

(a) *For all* $t \in \mathbb{N}$, X_t *is nonempty and convex,* $\hat{x}_t \in \operatorname{int} X_t$, *and* U_t *is nonempty.*
(b) *For all* $t \in \mathbb{N}$, *the partial functions* $\phi_t(., \hat{u}_t)$ *and* $f_t(., \hat{u}_t)$ *are continuous on a neighborhood of* $\hat{x}_t$ *and differentiable at* $\hat{x}_t$.
(c) *For all* $t \in \mathbb{N}$, *for all* $(x_t, x_{t+1}) \in X_t \times X_{t+1}$, $co A_t(x_t, x_{t+1}) \subset B_t(x_t, x_{t+1})$.
(d) *For all* $t \in \mathbb{N}$, *for all* $i, j \in \{1, \dots, n\}$, $\frac{\partial f^i(\hat{x}_t, \hat{u}_t)}{\partial x_t^j} \geq 0$, *and for all* $j \in \{1, \dots, n\}$, $\frac{\partial f^i(\hat{x}_t, \hat{u}_t)}{\partial x_t^j} > 0$.
(e) *For all* $t \in \mathbb{N}$, $\phi_t \geq 0$.
(f) *For all* $t \in \mathbb{N}$, *for all* $(y_{t+1}, y_t, u_t) \in X_{t+1} \times X_t \times U_t$ *such that* $y_{t+1} \leq f_t(y_t, u_t)$, *there exists* $v_t \in U_t$ *such that* $v_t \geq u_t$ *and* $y_{t+1} = f_t(y_t, v_t)$.
(g) *For all* $t \in \mathbb{N}$, *for all* $x_t \in X_t$, *the partial function* $\phi_t(x_t, .)$ *is increasing.*
(h) *For all* $t \in \mathbb{N}$, *for all* $z_t \in X_t$, *there exists* $s \in \mathbb{N}_*$ *and there exists* $(v_t, \dots, v_{t+s-1}) \in \prod_{j=0}^{s-1} U_{t+j}$ *such that by setting* $z_{t+j+1} := f_{t+j}(z_{t+j}, v_{t+j})$ *for* $j \in \{0, \dots, s-1\}$ *we have* $z_{t+s} = \hat{x}_{t+s}$.

Then there exist $\lambda_0 \in \mathbb{R}$, $(p_{t+1})_{t \in \mathbb{N}} \in (\mathbb{R}^{n*})^N$ *which satisfy the following conditions:*

(i) $(\lambda_0, p_1) \neq (0, 0)$.
(ii) $\lambda_0 \geq 0$.
(iii) *For all* $t \in \mathbb{N}$, $p_{t+1} \geq 0$ *and* $\langle p_{t+1}, f_t(\hat{x}_t, \hat{u}_t) - \hat{x}_{t+1} \rangle = 0$.
(iv) *For all* $t \in \mathbb{N}_*$, $p_t = D_1 H_t(\hat{x}_t, \hat{u}_t, p_{t+1}, \lambda_0)$.
(v) *For all* $t \in \mathbb{N}$, $H_t(\hat{x}_t, \hat{u}_t, p_{t+1}, \lambda_0^T) = \max_{u \in U_t} H_t(\hat{x}_t, u, p_{t+1}, \lambda_0)$.

Proof. Note that (CA 3) = (e), (CA 4) = (f), (CA 5) = (g), and (CA, $(\underline{\hat{x}}, \underline{\hat{u}})$) = (h). Then we can use Theorem 1.4 to assert that $(\underline{\hat{x}}, \underline{\hat{u}})$ is also a solution of $(\mathscr{P}_i^n)$. We conclude by using Theorem 2.11. □

Remark 2.2. These theorems, from Theorem 2.11 until Theorem 2.15, appear in the paper of Blot [11]. It is useful to note that in Theorems 2.12–2.15 the adjoint variables p_{t+1} are positive although the problem is governed by (DE).

In all the results of this subsection, we have used the condition of Michel. If we use the condition of Ioffe and Tihomirov, we obtain the following result which is new.

Theorem 2.16. *Let $(\underline{\hat{x}}, \underline{\hat{u}})$ be a solution of $(\mathscr{P}_i^n)$, or of $(\mathscr{P}_i^s)$, or of $(\mathscr{P}_i^o)$, or of $(\mathscr{P}_i^w)$. We assume that the following conditions are fulfilled:*

(a) For all $t \in \mathbb{N}$, $\phi_t(., \hat{u}_t)$ and $f_t(., \hat{u}_t)$ are of class C^1 at $\hat{x}$.
(b) For all $t \in \mathbb{N}$, there exists a neighborhood V_t of $\hat{x}_t$ in X_t such that, for all $x \in V_t$, for all $u_1, u_2 \in U_t$, for all $\theta \in [0, 1]$, there exists $u_3 \in U_t$ such that

$$\begin{cases} \phi_t(x, u_3) \geq (1-\theta)\phi_t(x, u_1) + \theta\phi_t(x, u_2) \\ f_t(x, u_3) \geq (1-\theta) f_t(x, u_1) + \theta f_t(x, u_2). \end{cases}$$

(c) For all $t \in \mathbb{N}$, for all $i, j \in \{1, \ldots, n\}$, $\frac{\partial f^i(\hat{x}_t, \hat{u}_t)}{\partial x_t^j} \geq 0$, and for all $j \in \{1, \ldots, n\}$, $\frac{\partial f^i(\hat{x}_t, \hat{u}_t)}{\partial x_t^j} > 0$.

Then there exist $\lambda_0 \in \mathbb{R}$, $(p_{t+1})_{t\in\mathbb{N}} \in (\mathbb{R}^{n*})^{\mathbb{N}}$ which satisfy the following conditions:

(i) $(\lambda_0, p_1) \neq (0, 0)$.
(ii) $\lambda_0 \geq 0$.
(iii) For all $t \in \mathbb{N}$, $p_{t+1} \geq 0$ and $\langle p_{t+1}, f_t(\hat{x}_t, \hat{u}_t) - \hat{x}_{t+1}\rangle = 0$.
(iv) For all $t \in \mathbb{N}_*$, $p_t = D_1 H_t(\hat{x}_t, \hat{u}_t, p_{t+1}, \lambda_0)$.
(v) For all $t \in \mathbb{N}$, $H_t(\hat{x}_t, \hat{u}_t, p_{t+1}, \lambda_0^T) = \max_{u \in U_t} H_t(\hat{x}_t, u, p_{t+1}, \lambda_0)$.

Proof. Using Propositions 1.2 and 1.10, we obtain, for all $T \in \mathbb{N}$, $T \geq 2$, $\lambda_0^T \in \mathbb{R}$ and $p_1^T, \ldots, p_T^T \in \mathbb{R}^{n*}$ which satisfy the conclusion of Proposition 1.10. Then we conclude as in the proof of Theorem 2.11. □

Remark 2.3. Proceeding as we do to establish the results from Theorem 2.12 until Theorem 2.15, we can obtain strong Pontryagin principles for the problems $(\mathscr{P}_e^o)$ and $(\mathscr{P}_e^n)$ where the part of the assumption which comes from the result of Michel is replaced by assumptions which come from Ioffe and Tihomirov.

2.3.3 A Condition of Partial Submersion

To avoid the invertibility condition, beside the positivity condition, we introduce another condition on the vector field of the dynamical system.

In this subsection, $\mathbb{R}^n$ is endowed with its usual inner product which is denoted by $(. \mid .)$. Following [81] (p. 410), when E and F are two Hilbert spaces, and when $T \in \mathscr{L}(E, F)$, the adjoint of T is $T^* \in \mathscr{L}(F, E)$ characterized by $(T.x \mid y) = (x \mid T^*.y)$. And so, in our problems, we will use $D_1 f_t(\hat{x}_t, \hat{u}_t)^* \in \mathscr{L}(\mathbb{R}^n, \mathbb{R}^*)$. When $\pi \in \mathbb{R}^{n*}$, we associate to π the vector $\pi^* \in \mathbb{R}^n$ characterized by $(\pi^* \mid y) = \langle \pi, y \rangle$ for all $y \in \mathbb{R}^n$. When $\pi \in \mathbb{R}^{n*}$ and $L \in \mathscr{L}(\mathbb{R}^n, \mathbb{R}^n)$, for all $y \in \mathbb{R}^n$, we have $\langle \pi \circ L, y \rangle = \pi(L.y) = \langle \pi, L.y \rangle = (\pi^* \mid L.y) = (L^*.\pi^* \mid y)$. Also recall that the gradient of a real-valued differentiable function is the vector in the primal space which represents the differential which belongs to the dual space. And so, in our problems, we will use of the partial gradient of ϕ_t, $(\nabla_1 \phi_t(x_t, u_t) \mid y) = \langle D_1 \phi_t(x_t, u_t), y \rangle$ for all $y \in \mathbb{R}^n$. And so, using these notions, the adjoint equation can be rewritten as

$$p_t^* = D_1 f_t(\hat{x}_t, \hat{u}_t)^*.p_{t+1}^* + \lambda_0 \nabla_1 \phi_t(\hat{x}_t, \hat{u}_t) \tag{2.33}$$

and the Hamiltonian can be written as

$$H_t(x_t, u_t, p_{t+1}, \lambda_0) = \lambda_0 \phi_t(x_t, u_t) + (p_{t+1}^* \mid f_t(x_t, u_t)). \tag{2.34}$$

Assuming the existence of the partial differential with respect to the state variable, we introduce the two following subspaces:

$$\left.\begin{array}{l} M_t := \mathrm{Im} D_1 f_t(\hat{x}_t, \hat{u}_t) \\ N_t := M_t^{\perp} = \mathrm{Ker} D_1 f_t(\hat{x}_t, \hat{u}_t)^* \end{array}\right\} \tag{2.35}$$

where $\perp$ denotes the orthogonal subspace. π_{M_t} and π_{N_t} denote the orthogonal projectors on M_t and on N_t. We also use the notation $S_{N_t}(x, \rho) := \{z \in N_t : \|z - x\| = \rho\}$ and $B_{N_t}(x, \rho) := \{z \in N_t : \|z - x\| \leq \rho\}$.

Now we can introduce our new condition.

$$\left.\begin{array}{l} \forall t \in \mathbb{N}, \exists P_t \subset U_t, P_t \neq \emptyset \text{ s.t.} \\ (\alpha)\ \exists \varrho_t > 0, \pi_{N_t}(f_t(\{\hat{x}_t\} \times P_t)) \supset (S_{N_t}(0, \varrho_t) + \pi_{N_t}(f_t(\hat{x}_t, \hat{u}_t)) \\ (\beta)\ \pi_{M_t}(f_t(\{\hat{x}_t\} \times P_t)) \text{ is bounded} \\ (\gamma)\ \phi_t(\{\hat{x}_t\} \times P_t) \text{ is bounded.} \end{array}\right\} \tag{2.36}$$

We also consider another condition which is simpler than (2.36).

$$\left.\begin{array}{l}\forall t \in \mathbb{N},\\ (1)\ \phi_t(\hat{x}_t, .)\ \text{and}\ f_t(\hat{x}_t, .)\ \text{are continuous on}\ U_t\\ (2)\ D_1 f_t(\hat{x}_t, \hat{u}_t)\ \text{exists}\\ (3)\ \hat{u}_t \in \operatorname{int} U_t\\ (4)\ f_t(\hat{x}_t, .)\ \text{is of class}\ C^1\ \text{at}\ \hat{u}_t\\ (5)\ \operatorname{Im}\pi_{N_t} \circ D_2 f_t(\hat{x}_t, \hat{u}_t) = N_t\end{array}\right\} \tag{2.37}$$

Remark 2.4. The condition (2.37) implies the condition (2.36). To justify that, note that, using condition (5), since $D_2(\pi_{N_t} \circ f_t)(\hat{x}_t, \hat{u}_t) = \pi_{N_t} \circ D_2 f_t(\hat{x}_t, \hat{u}_t)$ is surjective from $\mathbb{R}^d$ onto N_t, using a theorem of Graves ([64] p. 397), there exists a closed ball $P_t = \{u \in U_t : \|u - \hat{u}_t\| \leq r_t\}$ such that

$$\pi_{N_t} \circ f_t(\{\hat{x}_t\} \times P_t) \supset B_{N_t}(f_t(\hat{x}_t, \hat{u}_t), \varrho_t) \supset S_{N_t}(0, \varrho_t) + f_t(\hat{x}_t, \hat{u}_t),$$

and so the condition (α) of (2.36) is fulfilled. Since $\dim\mathbb{R}^d < +\infty$, P_t is compact. The continuities in condition (1) imply the conditions (β) and (γ).

Theorem 2.17. *Let* $(\underline{\hat{x}}, \underline{\hat{u}})$ *be a solution of* $(\mathcal{P}_a^n)$*, or of* $(\mathcal{P}_a^s)$*, or of* $(\mathcal{P}_a^o)$*, or of* $(\mathcal{P}_a^w)$ *when* $a \in \{e, i\}$*. We assume that the following conditions are fulfilled:*

(a) For all $t \in \mathbb{N}$*,* X_t *is a nonempty open convex subset of* $\mathbb{R}^n$ *and* U_t *is a nonempty subset of* $\mathbb{R}^d$*.*
(b) For all $t \in \mathbb{N}$*, the functions* ϕ_t *and* f_t *are differentiable with respect to the first vector variable.*
(c) For all $t \in \mathbb{N}$*, for all* $(x_t, x_{t+1}) \in X_t \times X_{t+1}$*,* $coA_t'(x_t, x_{t+1}) \subset B_t'(x_t, x_{t+1})$*.*
(d) Condition (2.36) holds.

Then there exist $\lambda_0 \in \mathbb{R}$ *and* $(p_{t+1})_{t\in\mathbb{N}} \in (\mathbb{R}^{n*})^{\mathbb{N}}$ *which satisfy the following conditions:*

(i) $(\lambda_0, p_1) \neq (0, 0)$*.*
(ii) $\lambda_0 \geq 0$*.*
(iii) For all $t \in \mathbb{N}$*,* $p_{t+1} \geq 0$ *and* $\langle p_{t+1}, f_t(\hat{x}_t, \hat{u}_t) - \hat{x}_{t+1}\rangle = 0$ *when* $a = i$*.*
(iv) For all $t \in \mathbb{N}_*$*,* $p_t = D_1 H_t(\hat{x}_t, \hat{u}_t, p_{t+1}, \lambda_0)$*.*
(v) For all $t \in \mathbb{N}$*,* $H_t(\hat{x}_t, \hat{u}_t, p_{t+1}, \lambda_0) = \max_{u \in U_t} H_t(\hat{x}_t, u, p_{t+1}, \lambda_0)$*.*

Proof. **The case a = e.** Using Propositions 1.2 and 1.7, we can assert that, for all $T \in \mathbb{N}$, $T \geq 2$, there exist $\lambda_0^T \in \mathbb{R}$ and $p_1^T, \ldots, p_T^T \in \mathbb{R}^{n*}$ which satisfy the conclusions of Proposition 1.8.

Since M_t is the orthogonal to $\operatorname{Ker} D_1 f_t(\hat{x}_t, \hat{u}_t)^*$, for all $z \in M_t$, we have $D_1 f_t(\hat{x}_t, \hat{u}_t)^*.z \neq 0$. Using the compactness of the unit sphere of M_t and the continuity of $D_1 f_t(\hat{x}_t, \hat{u}_t)^*$, we have $a_t := \inf\{\|D_1 f_t(\hat{x}_t, \hat{u}_t)^*.z\| : z \in M_t, \|z\| = 1\} > 0$. And so we have

$$\exists a_t \in (0, +\infty), \forall z \in M_t, \|D_1 f_t(\hat{x}_t, \hat{u}_t)^*.z\| \geq a_t.\|z\|. \tag{2.38}$$

Using the vector translation (2.33) of the third conclusion of Proposition 1.7, we obtain

$$\begin{aligned} p_t^{T*} &= D_1 f_t(\hat{x}_t, \hat{u}_t)^*.p_{t+1}^{T*} + \lambda_0^T \nabla_1 \phi_t(\hat{x}_t, \hat{u}_t) \\ &= D_1 f_t(\hat{x}_t, \hat{u}_t)^*.\pi_{M_t}(p_{t+1}^{T*}) + D_1 f_t(\hat{x}_t, \hat{u}_t)^*.\pi_{N_t}(p_{t+1}^{T*}) + \lambda_0^T \nabla_1 \phi_t(\hat{x}_t, \hat{u}_t) \\ &= D_1 f_t(\hat{x}_t, \hat{u}_t)^*.\pi_{M_t}(p_{t+1}^{T*}) + \lambda_0^T \nabla_1 \phi_t(\hat{x}_t, \hat{u}_t) \end{aligned}$$

which implies $p_t^{T*} - \lambda_0^T \nabla_1 \phi_t(\hat{x}_t, \hat{u}_t) = D_1 f_t(\hat{x}_t, \hat{u}_t)^*.\pi_{M_t}(p_{t+1}^{T*})$, and therefore, using (2.38), we obtain

$$\begin{aligned} \|p_t^{T*}\| + \lambda_0^T \|\nabla_1 \phi_t(\hat{x}_t, \hat{u}_t)\| &\geq \|p_t^{T*} - \lambda_0^T \nabla_1 \phi_t(\hat{x}_t, \hat{u}_t)\| \\ = \|D_1 f_t(\hat{x}_t, \hat{u}_t)^*.\pi_{M_t}(p_{t+1}^{T*})\| &\geq a_t.\|\pi_{M_t}(p_{t+1}^{T*})\| \end{aligned}$$

from which we have

$$\forall T > t, \qquad \|\pi_{M_t}(p_{t+1}^{T*})\| \leq \frac{1}{a_t}\|p_t^{T*}\| + \lambda_0^T \frac{1}{a_t}\|\nabla_1 \phi_t(\hat{x}_t, \hat{u}_t)\|. \tag{2.39}$$

Now we introduce the following notation:

$$\begin{cases} \Delta\phi_t(u_t) := \phi_t(\hat{x}_t, \hat{u}_t) - \phi_t(\hat{x}_t, u_t) \\ \Delta f_t(u_t) := f_t(\hat{x}_t, \hat{u}_t) - f_t(\hat{x}_t, u_t). \end{cases}$$

Using (2.34), the fourth conclusion of Proposition 1.7 implies, for all $u_t \in U_t$, $\lambda_0^T \Delta\phi_t(u_t) + (p_{t+1}^{T*} \mid \Delta f_t(u_t)) \geq 0$, which implies by using the orthogonality between M_t and N_t,

$$\lambda_0^T \Delta\phi_t(u_t) + (\pi_{M_t}(p_{t+1}^{T*}) \mid \pi_{M_t}(\Delta f_t(u_t))) + (\pi_{N_t}(p_{t+1}^{T*}) \mid \pi_{N_t}(\Delta f_t(u_t))) \geq 0$$

which implies

$$\begin{cases} \lambda_0^T \Delta\phi_t(u_t) + (\pi_{M_t}(p_{t+1}^{T*}) \mid \pi_{M_t}(\Delta f_t(u_t))) \\ \geq (\pi_{N_t}(p_{t+1}^{T*}) \mid \pi_{N_t}(f_t(\hat{x}_t, u_t))) - (\pi_{N_t}(p_{t+1}^{T*}) \mid \pi_{N_t}(f_t(\hat{x}_t, \hat{u}_t))). \end{cases}$$

Using the Cauchy–Schwarz–Buniakovski inequality, we obtain

$$\begin{cases} \lambda_0^T |\Delta\phi_t(u_t)| + \|\pi_{M_t}(p_{t+1}^{T*})\|.\|\pi_{M_t}(\Delta f_t(u_t))\| \\ \geq (\pi_{N_t}(p_{t+1}^{T*}) \mid \pi_{N_t}(f_t(\hat{x}_t, u_t))) - (\pi_{N_t}(p_{t+1}^{T*}) \mid \pi_{N_t}(f_t(\hat{x}_t, \hat{u}_t))). \end{cases}$$

Using conditions (β) and (γ) of the assumption (2.36) and the fact that the norm of an orthogonal projector is less than 1, we know that

$$\xi_t := \sup_{u_t \in U_t} |\Delta\phi_t(u_t)| < +\infty, \qquad \zeta_t := \sup_{u_t \in U_t} \|\pi_{M_t}(\Delta f_t(u_t))\| < +\infty.$$

And then using the previous inequalities, we obtain by taking the sup on the $u_t \in U_t$,

$$\begin{aligned}
&\lambda_0^T.\xi_t + \zeta_t.\|\pi_{M_t}(p_{t+1}^{T*})\| \\
&\quad \geq \sup_{u_t \in U_t} (\pi_{N_t}(p_{t+1}^{T*}) \mid \pi_{N_t}(f_t(\hat{x}_t, u_t))) - (\pi_{N_t}(p_{t+1}^{T*}) \mid \pi_{N_t}(f_t(\hat{x}_t, \hat{u}_t))) \\
&\quad \geq \sup_{z_t \in S_{N_t}(0,\varrho_t)} (\pi_{N_t}(p_{t+1}^{T*}) \mid z_t + \pi_{N_t}(f_t(\hat{x}_t, \hat{u}_t))) - (\pi_{N_t}(p_{t+1}^{T*}) \mid \pi_{N_t}(f_t(\hat{x}_t, \hat{u}_t))) \\
&\quad = \sup_{z_t \in S_{N_t}(0,\varrho_t)} (\pi_{N_t}(p_{t+1}^{T*}) \mid z_t) + (\pi_{N_t}(p_{t+1}^{T*}) \mid \pi_{N_t}(f_t(\hat{x}_t, \hat{u}_t))) \\
&\qquad - (\pi_{N_t}(p_{t+1}^{T*}) \mid \pi_{N_t}(f_t(\hat{x}_t, \hat{u}_t))) \\
&\quad = \sup_{z_t \in S_{N_t}(0,\varrho_t)} (\pi_{N_t}(p_{t+1}^{T*}) \mid z_t) \\
&\quad = \varrho_t. \sup_{w_t \in S_{N_t}(0,1)} (\pi_{N_t}(p_{t+1}^{T*}) \mid w_t) \\
&\quad = \varrho_t.\|\pi_{N_t}(p_{t+1}^{T*})\|,
\end{aligned}$$

and so we have proven the following property:

$$\forall T > t, \|\pi_{N_t}(p_{t+1}^{T*})\| \leq \frac{\xi_t}{\varrho_t}\lambda_0^T + \frac{\zeta_t}{\varrho_t}\|\pi_{M_t}(p_{t+1}^{T*})\|. \tag{2.40}$$

Using (2.39) in (2.40), we obtain the following inequalities, for all $T > t$:

$$\begin{cases}
\|\pi_{M_t}(p_{t+1}^{T*})\| \leq \dfrac{\|\nabla_1\phi_t(\hat{x}_t, \hat{u}_t)\|}{a_t}\lambda_0^T + \frac{1}{a_t}\|p_t^{T*}\| \\
\|\pi_{N_t}(p_{t+1}^{T*})\| \leq \left(\dfrac{\xi_t}{\varrho_t} + \dfrac{\zeta_t.\|\nabla_1\phi_t(\hat{x}_t, \hat{u}_t)\|}{\varrho_t.a_t}\right)\lambda_0^T + \dfrac{\zeta_t}{\varrho_t.a_t}\|p_t^{T*}\|,
\end{cases}$$

from which we deduce

$$\begin{aligned}
&\|p_{t+1}^T\|_* = \|p_{t+1}^{T*}\| = \|\pi_{M_t}(p_{t+1}^{T*}) + \pi_{N_t}(p_{t+1}^{T*})\| = \|\pi_{M_t}(p_{t+1}^{T*})\| + \|\pi_{N_t}(p_{t+1}^{T*})\| \\
&\leq \left(\frac{\xi_t}{\varrho_t} + \frac{(\zeta_t + \varrho_t)\|\nabla_1\phi_t(\hat{x}_t, \hat{u}_t)\|}{\varrho_t.a_t}\right)\lambda_0^T + \frac{\zeta_t + \varrho_t}{\varrho_t.a_t}\|p_t^{T*}\|
\end{aligned}$$

and so using the normalization $\|(\lambda_0^T, p_1^T)\| = 1$, from the previous inequality, by induction we obtain that, for all $t \in \mathbb{N}$, the sequence $T \mapsto p_t^T$ is bounded, and we can conclude as in the proof of Theorem 2.1.

The case a = i. The reasoning is similar using Proposition 1.8 instead of Proposition 1.7. □

To finish this subsection, we use the condition of Ioffe and Tihomirov.

Theorem 2.18. *Let $(\underline{\hat{x}}, \underline{\hat{u}})$ be a solution of $(\mathcal{P}_i^n)$, or of $(\mathcal{P}_i^s)$, or of $(\mathcal{P}_i^o)$, or of $(\mathcal{P}_i^w)$. We assume that the following conditions are fulfilled:*

(a) For all $t \in \mathbb{N}$, $\phi_t(., \hat{u}_t)$ and $f_t(., \hat{u}_t)$ are of class C^1 at $\hat{x}$.

(b) For all $t \in \mathbb{N}$, there exists a neighborhood V_t of $\hat{x}_t$ in X_t such that, for all $x \in V_t$, for all $u_1, u_2 \in U_t$, for all $\theta \in [0,1]$, there exists $u_3 \in U_t$ such that

$$\begin{cases} \phi_t(x, u_3) \geq (1-\theta)\phi_t(x, u_1) + \theta\phi_t(x, u_2) \\ f_t(x, u_3) \geq (1-\theta) f_t(x, u_1) + \theta f_t(x, u_2). \end{cases}$$

(c) The condition (2.36) holds.

Then there exist $\lambda_0 \in \mathbb{R}$, $(p_{t+1})_{t \in \mathbb{N}} \in (\mathbb{R}^{n*})^N$ which satisfy the following conditions:

(i) $(\lambda_0, p_1) \neq (0,0)$.
(ii) $\lambda_0 \geq 0$.
(iii) For all $t \in \mathbb{N}$, $p_{t+1} \geq 0$ and $\langle p_{t+1}, f_t(\hat{x}_t, \hat{u}_t) - \hat{x}_{t+1} \rangle = 0$.
(iv) For all $t \in \mathbb{N}_*$, $p_t = D_1 H_t(\hat{x}_t, \hat{u}_t, p_{t+1}, \lambda_0)$.
(v) For all $t \in \mathbb{N}$, $H_t(\hat{x}_t, \hat{u}_t, p_{t+1}, \lambda_0^T) = \max_{u \in U_t} H_t(\hat{x}_t, u, p_{t+1}, \lambda_0)$.

Proof. Using Propositions 1.2 and 1.9 when $a = e$ or Proposition 1.10 when $a = i$, we obtain $\lambda_0^T \in \mathbb{R}$, $p_1^T, \ldots, p_T^T \in \mathbb{R}^{n*}$ which satisfy the conclusions of Proposition 1.9 when $a = e$ or of Proposition 1.10 when $a = i$. And then we conclude as in the proof of Theorem 2.17. □

Remark 2.5. Proceeding as we do to establish the results from Theorem 2.12 until Theorem 2.15, we can obtain strong Pontryagin principles for the problems $(\mathscr{P}_e^o)$ and for $(\mathscr{P}_e^n)$ where the part of the assumption which comes from the result of Michel is replaced by assumptions which come from Ioffe and Tihomirov.

2.4 Constrained Problems

In this section we still consider systems governed by (DE) or (DI). We consider constraints which possess the following form, for all $t \in \mathbb{N}$, when $x_t \in X_t$:

$$\begin{aligned} \mathscr{U}_t(x_t) := \{u_t \in U_t : \forall j \in \{1, \ldots, d^i\}, g_t^j(x_t, u_t) \\ \geq 0, \forall k \in \{1, \ldots, d^e\}, h_t^k(x_t, u_t) = 0\}. \end{aligned} \tag{2.41}$$

The terminology varies when we speak of such constraints. Following [6] (p. 221) these constraints represent a "feedback perfect state information": the value of the state variable x_t modifies the set of all admissible values of the control variable u_t. We define the admissible processes which satisfy these constraints, when $a \in \{e, i\}$.

$$\mathrm{Adm}_{\eta,c}^a := \{(\underline{x}, \underline{u}) \in \mathrm{Adm}_\eta^a : \forall t \in \mathbb{N}, u_t \in \mathscr{U}_t(x_t)\}. \tag{2.42}$$

We define the problems where these constraints are present, when $a \in \{e, i\}$.

$(\mathscr{C}_a^n)$ Maximize $J(\underline{x}, \underline{u})$ when $(\underline{x}, \underline{u}) \in \mathrm{Dom}_\eta^a(J) \cap \mathrm{Adm}_{\eta,c}^a$.

$(\mathscr{C}_a^s)$ Find $(\underline{\hat{x}}, \underline{\hat{u}}) \in \mathrm{Dom}_\eta^a(J) \cap \mathrm{Adm}_{\eta,c}^a$ such that, for all $(\underline{x}, \underline{u}) \in \mathrm{Adm}_{\eta,c}^a$,

$$J(\underline{\hat{x}}, \underline{\hat{u}}) \geq \limsup_{T\to+\infty} \sum_{t=0}^{T} \phi_t(x_t, u_t).$$

$(\mathscr{C}_a^o)$ Find $(\underline{\hat{x}}, \underline{\hat{u}}) \in \mathrm{Adm}_{\eta,c}^a$ such that, for all $(\underline{x}, \underline{u}) \in \mathrm{Adm}_{\eta,c}^a$,

$$\liminf_{T\to+\infty} \sum_{t=0}^{T} (\phi_t(\hat{x}_t, \hat{u}_t) - \phi_t(x_t, u_t)) \geq 0.$$

$(\mathscr{C}_a^w)$ Find $(\underline{\hat{x}}, \underline{\hat{u}}) \in \mathrm{Adm}_{\eta,c}^a$ such that, for all $(\underline{x}, \underline{u}) \in \mathrm{Adm}_{\eta,c}^a$,

$$\limsup_{T\to+\infty} \sum_{t=0}^{T} (\phi_t(\hat{x}_t, \hat{u}_t) - \phi_t(x_t, u_t)) \geq 0.$$

Besides the Hamiltonian H_t defined in Chap. 1, we consider the Lagrangian $L_t : X_t \times U_t \times \mathbb{R}^{n*} \times \mathbb{R} \times \mathbb{R}^{d^i *} \times \mathbb{R}^{d^e *} \to \mathbb{R}$ by setting

$$L_t(x, u, p, \lambda, \mu, \nu) := H_t(x, u, p, \lambda) + \langle \mu, g_t(x, u) \rangle + \langle \nu, h_t(x, u) \rangle. \tag{2.43}$$

where $g_t := (g_t^1, \ldots, g_t^{d^i})$ and $h_t := (h_t^1, \ldots h_t^{d^e})$.

Theorem 2.19. *Let $(\underline{\hat{x}}, \underline{\hat{u}})$ be a solution of $(\mathscr{C}_a^n)$, or of $(\mathscr{C}_a^s)$, or of $(\mathscr{C}_a^o)$, or of $(\mathscr{C}_a^w)$ where $a \in \{e, i\}$. We assume that the following conditions are fulfilled:*

(1) For all $t \in \mathbb{N}$, X_t is nonempty open and convex, and the functions ϕ_t, f_t, g_t, h_t are continuous on a neighborhood of $(\hat{x}_t, \hat{u}_t)$ and differentiable at $(\hat{x}_t, \hat{u}_t)$.

(2) For all $t \in \mathbb{N}$, $D_1 f_t(\hat{x}_t, \hat{u}_t)$ is invertible.

(3) Setting $S_t^j := D_1 g_t^j(\hat{x}_t, \hat{u}_t) \circ (D_1 f_t(\hat{x}_t, \hat{u}_t))^{-1} \circ D_2 f_t(\hat{x}_t, \hat{u}_t) - D_2 g_t^j(\hat{x}_t, \hat{u}_t)$ and $M_t^k := D_1 h_t^k(\hat{x}_t, \hat{u}_t) \circ (D_1 f_t(\hat{x}_t, \hat{u}_t))^{-1} \circ D_2 f_t(\hat{x}_t, \hat{u}_t) - D_2 h_t^k(\hat{x}_t, \hat{u}_t)$ for all $t \in \mathbb{N}$, the family $((S_t^j)_{1 \leq j \leq d^i}, (M_t^k)_{1 \leq k \leq d^e})$ is linearly independent.

Then there exist $\lambda_0 \in \mathbb{R}$, $(p_{t+1})_{t\in\mathbb{N}} \in (\mathbb{R}^{n*})^{\mathbb{N}}$, $(\mu_t)_{t\in\mathbb{N}} \in (\mathbb{R}^{d^i *})^{\mathbb{N}}$ and $(\nu_t)_{t\in\mathbb{N}} \in (\mathbb{R}^{d^e *})^{\mathbb{N}}$ which satisfy the following conditions:

(i) $(\lambda_0, p_1, \mu_0, \nu_0) \neq (0, 0, 0, 0)$.

(ii) $\lambda_0 \geq 0$, $\mu_t \geq 0$ and $\langle \mu_t, g_t(\hat{x}_t, \hat{u}_t) \rangle = 0$ for all $t \in \mathbb{N}$.

(iii) For all $t \in \mathbb{N}$, $p_{t+1} \geq 0$ and $\langle p_{t+1}, f_t(\hat{x}_t, \hat{u}_t) - \hat{x}_{t+1} \rangle = 0$ when $a = i$.

(iv) For all $t \in \mathbb{N}$, $p_t = D_1 L_t(\hat{x}_t, \hat{u}_t, p_{t+1}, \lambda_0, \mu_t, \nu_t)$.

(v) For all $t \in \mathbb{N}$, $D_2 L_t(\hat{x}_t, \hat{u}_t, p_{t+1}, \lambda_0, \mu_t, \nu_t) = 0$.

Proof. We do the proof in the case $a = e$. The case $a = i$ is similar. We use the method of reduction to the finite horizon. For all $T \in \mathbb{N}$, $T \geq 2$, the restriction $(\hat{x}_0, \ldots, \hat{x}_T, \hat{u}_0, \ldots, \hat{u}_{T-1})$ is a solution of the problem

$$(\mathscr{F}C_e(T,\eta,\hat{x}_T)) \begin{cases} \text{maximize } J_T(x_0,\ldots,x_T,u_0,\ldots,u_{T-1}) \\ \text{when } \quad \forall t \in \{0,\ldots,T-1\}, x_{t+1} = f_t(x_t,u_t) \\ \qquad\qquad \forall t \in \{0,\ldots,T-1\}, u_t \in \mathscr{U}_t(x_t) \\ \qquad\qquad x_0 = \eta, x_T = \hat{x}_T. \end{cases}$$

Note that x_0 and x_T are not variables of this problem. As in Sect. 1.4, we translate this problem into a problem of static optimization on which we can use the multiplier rule of Halkin that permits to obtain $\lambda_0^T \in \mathbb{R}$, $p_1^T, \ldots, p_T^T \in \mathbb{R}^{n*}$, $\mu_0^T, \ldots, \mu_{T-1}^T \in \mathbb{R}^{d^i*}$, $\nu_0^T, \ldots, \nu_{T-1}^T \in \mathbb{R}^{d^e*}$ such that the following properties hold:

$$(\lambda_0^T, p_1^T, \ldots, p_T^T, \mu_0^T, \ldots, \mu_{T-1}^T, \nu_0^T, \ldots, \nu_{T-1}^T) \neq 0. \tag{2.44}$$

$$\lambda_0^T, \forall t \in \{0,\ldots,T-1\}, \mu_t^T \geq 0, \langle \mu_t^T, g_t(\hat{x}_t,\hat{u}_t)\rangle = 0. \tag{2.45}$$

$$\forall t \in \{1,\ldots,T-1\}, p_t^T = D_1 L_t(\hat{x}_t,\hat{u}_t,p_{t+1}^T,\lambda_0^T,\mu_t^T,\nu_t^T). \tag{2.46}$$

$$\forall t \in \mathbb{N}, D_2 L_t(\hat{x}_t,\hat{u}_t,p_{t+1}^T,\lambda_0^T,\mu_t^T,\nu_t^T) = 0. \tag{2.47}$$

We want to prove the following assertion:

$$(\lambda_0^T, p_1^T, \mu_0^T, \nu_0^T) \neq 0. \tag{2.48}$$

To abridge the writing, we set $\hat{\phi}_t := \phi_t(\hat{x}_t,\hat{u}_t)$, $\hat{f}_t := f_t(\hat{x}_t,\hat{u}_t)$, $\hat{g}_t := g_t(\hat{x}_t,\hat{u}_t)$, $\hat{h}_t := h_t(\hat{x}_t,\hat{u}_t)$. We proceed by contradiction, we assume that $(\lambda_0^T, p_1^T, \mu_0^T, \nu_0^T) = 0$. Then using (2.46) and (2.47) for $t = 1$ we obtain $0 = p_2^T \circ D_1\hat{f}_t + \mu_1^T \circ D_1\hat{g}_t + \nu_1^T \circ D_1\hat{h}_t$ and $0 = p_2^T \circ D_2\hat{f}_t + \mu_1^T \circ D_2\hat{g}_t + \nu_1^T \circ D_2\hat{h}_t$ from which we deduce $-p_2^T = \mu_1^T \circ D_1\hat{g}_t \circ (D_1\hat{f}_t)^{-1} + \nu_1^T \circ D_1\hat{h}_t \circ (D_1\hat{f}_t)^{-1}$ and $-p_2^T \circ D_2\hat{f}_t = \mu_1^T \circ D_2\hat{g}_t + \nu_1^T \circ D_2\hat{h}_t$, which implies

$$\begin{aligned} -p_2^T \circ D_2\hat{f}_t &= \mu_1^T \circ D_1\hat{g}_t \circ (D_1\hat{f}_t)^{-1} \circ D_2\hat{f}_t + \nu_1^T \circ D_1\hat{h}_t \circ (D_1\hat{f}_t)^{-1} \circ D_2\hat{f}_t \\ &= \mu_1^T \circ D_2\hat{g}_t + \nu_1^T \circ D_2\hat{h}_t \end{aligned}$$

Denoting $S_t := (S_t^1,\ldots,S_t^{d^i})$ and $M_t := (M_t^1,\ldots,M_t^{d^e})$, we deduce from the last relation

$$\mu_1^T \circ S_t + \nu_1^T \circ M_t = 0,$$

and using the coordinates and the assumption (3) we obtain $\mu_1^T = 0$ and $\nu_1^T = 0$. Then (2.46) for $t = 1$ implies $p_2^T \circ D_1\hat{f}_t = 0$, and assumption (2) implies $p_2^T = 0$. And so we have proven that $(\lambda_0^T, p_1^T, \mu_0^T, \nu_0^T) = 0$ implies $(\lambda_0^T, p_2^T, \mu_1^T, \nu_1^T) = 0$. Iterating this reasoning we obtain, for all $t \in \mathbb{N}$, $p_{t+1}^T = 0$, $\mu_t^T = 0$, and $\nu_t^T = 0$, which is a contradiction with (2.44). And so (2.48) is proven. Using a normalization, i.e., multiplying all the multipliers by $\|(\lambda_0^T, p_1^T, \mu_0^T, \nu_0^T)\|$, we can assume that $\|(\lambda_0^T, p_1^T, \mu_0^T, \nu_0^T)\| = 1$. Consequently the sequences $T \mapsto \lambda_0^T$,

$T \mapsto p_1^T$, $T \mapsto \mu_0^T$, and $T \mapsto \nu_0^T$ are bounded. Then from (2.46) we obtain $p_2^T = (p_1^T - \lambda_0^T D_1\hat{\phi}_1 - \mu_1^T \circ D_1\hat{g}_1 - \nu_1^T \circ D_1\hat{h}_1) \circ (D_1\hat{f}_1)^{-1}$ which implies $p_2^T \circ D_2\hat{f}_1 = (p_1^T - \lambda_0^T D_1\hat{\phi}_1 - \mu_1^T \circ D_1\hat{g}_1 - \nu_1^T \circ D_1\hat{h}_1) \circ (D_1\hat{f}_1)^{-1} \circ D_2\hat{f}_1$, and from (2.47) we obtain $p_2^T \circ D_2\hat{f}_1 = -\lambda_0^T D_2\hat{\phi}_1 - \mu_1^T \circ D_2\hat{g}_1 - \nu_1^T \circ D_2\hat{h}_1$. From these two equalities we deduce

$$\mu_1^T \circ S_1 + \nu_1^T \circ M_1 = p_1^T \circ (D_1\hat{f}_1)^{-1} \circ D_2\hat{f}_1 + \lambda_0^T (D_2\hat{\phi}_1 - D_1\hat{\phi}_1 \circ (D_1\hat{f}_1)^{-1} \circ D_2\hat{f}_1.$$

The right-hand term is bounded as function of T, and consequently $T \mapsto \mu_1^T \circ S_1 + \nu_1^T \circ M_1$ is bounded. Then, translating this expression in terms of coordinates, using assumption (3) and Lemma 2.2, we obtain that the sequences $T \mapsto \mu_1^T$ and $T \mapsto \nu_1^T$ are bounded. And then, from $p_2^T = (p_1^T - \lambda_0^T D_1\hat{\phi}_1 - \mu_1^T \circ D_1\hat{g}_1 - \nu_1^T \circ D_1\hat{h}_1) \circ (D_1\hat{f}_1)^{-1}$, we obtain that $T \mapsto p_2^T$ is bounded. Iterating this reasoning, we obtain that, for all $t \in \mathbb{N}$, the sequences $T \mapsto p_t^T$, $T \mapsto \mu_t^T$ and $T \mapsto \nu_t^T$ are bounded. And then we can use Lemma 2.1 and conclude as in the proof of Theorem 2.3. □

This result appears in the paper of Blot [12]. To finish this section we give a strong Pontryagin principle. We consider the following simplified constraints:

$$\mathscr{U}_t^1(x_t) := \{u_t \in U_t : \forall j \in \{1, \ldots, d^i\}, g_t^j(x_t, u_t) \geq 0\}. \tag{2.49}$$

For $\ell \in \{n, s, o, w\}$, we denote by $(\mathscr{C}_e^{\ell,1})$ the problem obtained by replacing $\mathscr{U}_t(x_t)$ by $\mathscr{U}_t^1(x_t)$ into $(\mathscr{C}_a^n)$. For these simplified constraints, the Lagrangian becomes $L_t^1(x, u, p, \lambda, \mu) := \lambda\phi_t(x, u) + \langle p, f_t(x, u)\rangle + \langle \mu, g_t(x, u)\rangle$. To use the condition of Michel, we ought to consider $A_t(x_t, x_{t+1})$ as the set of all $(r_t, \zeta_t, \xi_t) \in U_t \times \mathbb{R}^n \times \mathbb{R}^{d^i}$ for which there exists $u_t \in U_t$ satisfying $r_t \leq \phi_t(x_t, u_t)$, $\zeta_t = f_t(x_t, u_t) - x_{t+1}$ and $\xi_t \leq g_t(x_t, u_t)$. $B_t(x_t, x_{t+1})$ is the set of all $(r_t, \zeta_t, \xi_t) \in U_t \times \mathbb{R}^n \times \mathbb{R}^{d^i}$ for which there exists $(u_t, \alpha_t, \beta_t) \in U_t \times \mathbb{R}^n \times \mathbb{R}^{d^i}$ satisfying $r_t \leq \phi_t(x_t, u_t)$, $\alpha_t^k \zeta_t^k = f_t^k(x_t, u_t) - x_{t+1}^k$ for all $k \in \{1, \ldots, n\}$and $\beta_t^j \xi_t^j \leq g_t^j(x_t, u_t)$ for all $j \in \{1, \ldots, d^i\}$.

Theorem 2.20. *Let $(\underline{\hat{x}}, \underline{\hat{u}})$ be a solution of $(\mathscr{C}_e^{n,1})$, or of $(\mathscr{C}_e^{s,1})$, or of $(\mathscr{C}_e^{o,1})$, or of $(\mathscr{C}_e^{w})$. We assume that the following conditions are fulfilled:*

(1) For all $t \in \mathbb{N}$, X_t is nonempty open and convex, the functions ϕ_t, f_t, g_t are continuous on a neighborhood of $(\hat{x}_t, \hat{u}_t)$ and differentiable at $(\hat{x}_t, \hat{u}_t)$.
(2) For all $t \in \mathbb{N}$, $D_1 f_t(\hat{x}_t, \hat{u}_t)$ is invertible.
(3) For all $t \in \mathbb{N}$, for all $(x_t, x_{t+1}) \in X_t \times X_{t+1}$, $coA_t(x_t, x_{t+1}) \subset B_t(x_t, x_{t+1})$.
(4) For all $t \in \mathbb{N}$, there exists $\tilde{u}_t \in U_t(\hat{x}_t)$ such that $f_t(\hat{x}_t, \tilde{u}_t) = f_t(\hat{x}_t, \hat{u}_t)$ and $g_t^j(\hat{x}_t, \tilde{u}_t) > 0$, for all $j \in \{1, \ldots, d^i\}$.

Then there exist $\lambda_0 \in R$, $(p_{t+1})_{t\in\mathbb{N}} \in (\mathbb{R}^{n*})^{\mathbb{N}}$, $(\mu_t)_{t\in\mathbb{N}} \in (\mathbb{R}^{d^i}*)^{\mathbb{N}}$ which satisfy the following conditions:

(i) $(\lambda, p_1) \neq (0, 0)$.
(ii) $\lambda_0 \geq 0$, $\mu_t \geq 0$ and $\langle \mu_t, g_t(\hat{x}_t, \hat{u}_t)\rangle = 0$ for all $t \in \mathbb{N}$.

(iii) For all $t \in \mathbb{N}$, $p_t = D_1 L_t^1(\hat{x}_t, \hat{u}_t, p_{t+1}, \lambda_0, \mu_t)$.
(iv) For all $t \in \mathbb{N}$, $L_t^1(\hat{x}_t, \hat{u}_t, p_{t+1}, \lambda_0, \mu_t) = \max_{u_t \in U_t} L_t^1(\hat{x}_t, u_t, p_{t+1}, \lambda_0, \mu_t)$.

This result appears in the paper of Blot and Hayek [23] where a proof is given. This paper contains other results on the constrained problems.

2.5 Multiobjective Problems

Results of the previous sections are extended to multiobjective problems by using similar methods. All the results of this section are due to Hayek [51] and [50]. The controlled dynamical systems are still (DE) and (DI). The difference with the previous sections is that we replace ϕ_t by several functions $\phi_{1,t}, \ldots, \phi_{m,t}$ from $X_t \times U_t$ into $\mathbb{R}$. We define $J_j(\underline{x}, \underline{u}) := \sum_{t=0}^{+\infty} \phi_{j,t}(x_t, u_t)$ when the series converges in $\mathbb{R}$. And we define $\text{Dom}_\eta^a(J_j)$ as the set of all $(\underline{x}, \underline{u}) \in \text{Adm}_\eta$ such that the series $\sum_{t=0}^{+\infty} \phi_{j,t}(x_t, u_t)$ converges in $\mathbb{R}$. We introduce the notation $\text{Dom}_\eta^a((J_j)_{1 \leq j \leq m}) := \bigcap_{j=1}^{m} \text{Dom}_\eta^a(J_j)$. The notions of optimality are notions of Pareto optimality and of weak Pareto optimality. Precisely the considered problems are the following ones.

($\mathcal{V}_a^n$) Find $(\underline{\hat{x}}, \underline{\hat{u}}) \in \text{Dom}_\eta^a((J_j)_{1 \leq j \leq m})$ such that there does not exist any $(\underline{x}, \underline{u}) \in \text{Dom}_\eta^a((J_j)_{1 \leq j \leq m})$ such that $J_j(\underline{x}, \underline{u}) \geq J_j(\underline{\hat{x}}, \underline{\hat{u}})$ for all $j \in \{1, \ldots, m\}$ and $J_h(\underline{x}, \underline{u}) > J_h(\underline{\hat{x}}, \underline{\hat{u}})$ for some $h \in \{1, \ldots, m\}$.

($\mathcal{V}_a^{n,w}$) Find $(\underline{\hat{x}}, \underline{\hat{u}} \in \text{Dom}_\eta^a((J_j)_{1 \leq j \leq m})$ such that there does not exist any $(\underline{x}, \underline{u}) \in \text{Dom}_\eta^a((J_j)_{1 \leq j \leq m})$ such that $J_j(\underline{x}, \underline{u}) > J_j(\underline{\hat{x}}, \underline{\hat{u}})$ for all $j \in \{1, \ldots, m\}$.

($\mathcal{V}_a^o$) Find $(\underline{\hat{x}}, \underline{\hat{u}}) \in \text{Adm}_\eta^a$ such that there does not exist any $(\underline{x}, \underline{u}) \in \text{Adm}_\eta^a$ such that $\limsup_{T \to +\infty} \sum_{t=0}^{T} (\phi_{j,t}(x_t, u_t) - \phi_{j,t}(\hat{x}_t, \hat{u}_t)) \geq 0$ for all $j \in \{1, \ldots, m\}$ and $\limsup_{T \to +\infty} \sum_{t=0}^{T} (\phi_{h,t}(x_t, u_t) - \phi_{h,t}(\hat{x}_t, \hat{u}_t)) > 0$ for some $h \in \{1, \ldots, m\}$.

($\mathcal{V}_a^{o,w}$) Find $(\underline{\hat{x}}, \underline{\hat{u}}) \in \text{Adm}_\eta^a$ such that there does not exist any $(\underline{x}, \underline{u}) \in \text{Adm}_\eta^a$ such that $\limsup_{T \to +\infty} \sum_{t=0}^{T} (\phi_{j,t}(x_t, u_t) - \phi_{j,t}(\hat{x}_t, \hat{u}_t)) > 0$ for all $j \in \{1, \ldots, m\}$.

($\mathcal{V}_a^w$) Find $(\underline{\hat{x}}, \underline{\hat{u}}) \in \text{Adm}_\eta^a$ such that there does not exist any $(\underline{x}, \underline{u}) \in \text{Adm}_\eta^a$ such that $\liminf_{T \to +\infty} \sum_{t=0}^{T} (\phi_{j,t}(x_t, u_t) - \phi_{j,t}(\hat{x}_t, \hat{u}_t)) \geq 0$ for all $j \in \{1, \ldots, m\}$ and $\liminf_{T \to +\infty} \sum_{t=0}^{T} (\phi_{h,t}(x_t, u_t) - \phi_{h,t}(\hat{x}_t, \hat{u}_t)) > 0$ for some $h \in \{1, \ldots, m\}$.

$(\mathcal{V}_a^{w,w})$ Find $(\underline{\hat{x}}, \underline{\hat{u}}) \in \mathrm{Adm}_\eta^a$ such that there does not exist any $(\underline{x}, \underline{u}) \in \mathrm{Adm}_\eta^a$ such that $\liminf_{T \to +\infty} \sum_{t=0}^{T} (\phi_{j,t}(x_t, u_t) - \phi_{j,t}(\hat{x}_t, \hat{u}_t)) > 0$ for all $j \in \{1, \ldots, m\}$.

A solution of $(\mathcal{V}_a^n)$ (respectively $(\mathcal{V}_a^o)$, respectively $(\mathcal{V}_a^w)$) is called a Pareto optimal solution (respectively an overtaking Pareto optimal solution, respectively a weak overtaking Pareto optimal solution). For the solutions of $(\mathcal{V}_a^{n,w})$, $(\mathcal{V}_a^{o,w})$, $(\mathcal{V}_a^{w,w})$ we replace Pareto by weak Pareto optima.

We start with a first result of necessary conditions for weak Pareto optima in the form of a weak Pontryagin principle.

Theorem 2.21. *Let $(\underline{\hat{x}}, \underline{\hat{u}})$ be a solution of $(\mathcal{V}_a^{n,w})$, or of $(\mathcal{V}_a^{o,w})$, or of $(\mathcal{V}_a^{w,w})$ when $a \in \{e, i\}$. We assume that the following conditions are fulfilled:*

(a) For all $t \in \mathbb{N}$, $\hat{u}_t \in \operatorname{int} U_t$, $\phi_{j,t}$, and f_t are of class C^1 at $(\hat{x}_t, \hat{u}_t)$ for all $j \in \{1, \ldots, m\}$.
(b) For all $t \in \mathbb{N}$, $D_1 f_t(\hat{x}_t, \hat{u}_t)$ is invertible.

Then there exist $(\lambda_1, \ldots, \lambda_m) \in \mathbb{R}^m$ and $(p_{t+1})_{t \in \mathbb{N}} \in (\mathbb{R}^{n*})^{\mathbb{N}}$ which satisfy the following conditions:

(i) $(\lambda_1, \ldots, \lambda_m, p_1) \neq (0, \ldots, 0, 0)$.
(ii) For all $j \in \{1, \ldots, m\}$, $\lambda_j \geq 0$, and when $a = i$, for all $t \in \mathbb{N}$, $p_{t+1} \geq 0$ and $\langle p_{t+1}, f_t(\hat{x}_t, \hat{u}_t) - \hat{x}_{t+1} \rangle = 0$.
(iii) For all $t \in \mathbb{N}_*$, $p_t = \sum_{j=1}^{m} \lambda_j D_1 \phi_{j,t}(\hat{x}_t, \hat{u}_t) + p_{t+1} \circ D_1 f_t(\hat{x}_t, \hat{u}_t)$.
(iv) For all $t \in \mathbb{N}$, $\sum_{j=1}^{m} \lambda_j D_2 \phi_{j,t}(\hat{x}_t, \hat{u}_t) + p_{t+1} \circ D_2 f_t(\hat{x}_t, \hat{u}_t) = 0$.

The proof of this result uses the method of reduction to finite horizon. Since the associated finite-horizon problems are now multiobjective problems while they were single-objective problems in the previous sections, the multiplier rules of static optimization (of Halkin or Clarke) are replaced by a multiplier rule which is special to static multiobjective problems and based on a theorem of Motzkin [51]. After that, the question is to extract the multipliers of the infinite-horizon problem from the sequences of multipliers of the finite-horizon problems, and the reasoning is similar to the reasoning of the previous sections.

Remark 2.6. When $a = i$, there exists in [51] a theorem where the condition of invertibility is replaced by the positivity condition as defined in Sect. 2.1.2 for single-objective problems. Moreover in the previous theorem, if in addition we assume that $D_2 f_0(\eta, \hat{u}_0)$ is onto, we have $(\lambda_1, \ldots, \lambda_m) \neq (0, \ldots, 0)$.

After a weak Pontryagin principle, we state a result in the form of a strong Pontryagin principle.

Theorem 2.22. *Let $(\underline{\hat{x}}, \underline{\hat{u}})$ be a solution of $(\mathcal{V}_a^{n,w})$, or of $(\mathcal{V}_a^{o,w})$, or of $(\mathcal{V}_a^{w,w})$ when $a \in \{e, i\}$. We assume that the following conditions are fulfilled:*

(a) For all $t \in \mathbb{N}$, X_t is convex, and for all $j \in \{1, \ldots, m\}$, for all $u_t \in U_t$, $\phi_{j,t}(., u_t)$ and $f_t(., u_t)$ are of class C^1 at $\hat{x}_t$.
(b) For all $t \in \mathbb{N}$, $D_1 f_t(\hat{x}_t, \hat{u}_t)$ is invertible.
(c) For all $t \in \mathbb{N}$, for all $x_t \in X_t$, for all $u'_t, u''_t \in U_t$, for all $\theta \in [0,1]$, there exists $u_t \in U_t$ such that, for all $j \in \{1, \ldots, m\}$, $\phi_{j,t}(x_t, u_t) \geq (1-\theta)\phi_{j,t}(x_t, u'_t) + \theta\phi_{j,t}(x_t, u''_t)$ and $f_t(x_t, u_t) = (1-\theta) f_t(x_t, u'_t) + \theta f_t(x_t, u''_t)$.

Then there exist $(\lambda_1, \ldots, \lambda_m) \in \mathbb{R}^m$ and $(p_{t+1})_{t \in \mathbb{N}} \in (\mathbb{R}^{n*})^{\mathbb{N}}$ which satisfy the following conditions:

(i) $(\lambda_1, \ldots, \lambda_m, p_1) \neq (0, \ldots, 0, 0)$.
(ii) For all $j \in \{1, \ldots, m\}$, $\lambda_j \geq 0$, and when $a = i$, for all $t \in \mathbb{N}$, $p_{t+1} \geq 0$ and $\langle p_{t+1}, f_t(\hat{x}_t, \hat{u}_t) - \hat{x}_{t+1} \rangle = 0$.
(iii) For all $t \in \mathbb{N}_*$, $p_t = \sum_{j=1}^{m} \lambda_j D_1\phi_{j,t}(\hat{x}_t, \hat{u}_t) + p_{t+1} \circ D_1 f_t(\hat{x}_t, \hat{u}_t)$.
(iv) For all $t \in \mathbb{N}$, for all $u_t \in U_t$, $\sum_{j=1}^{m} \lambda_j \phi_{j,t}(\hat{x}_t, \hat{u}_t) + \langle p_{t+1}, f_t(\hat{x}_t, \hat{u}_t) \rangle \geq \sum_{j=1}^{m} \lambda_j \phi_{j,t}(\hat{x}_t, u_t) + \langle p_{t+1}, f_t(\hat{x}_t, u_t) \rangle$.

The proof of this result also uses the method of the reduction to finite horizon. The tool of static multiobjective optimization which is used is a theorem of Khanh and Nuong (Theorem 2.2 in [58]). We recognize in assumption (c) a generalization of the condition of Ioffe and Tihomirov. The end of the proof is similar to this one of strong Pontryagin principles of the previous sections.

Remark 2.7. If moreover we assume that $d \geq n$ and that $f_0(\{\eta\} \times U_0) - \hat{x}_1$ is a neighborhood of 0 in $\mathbb{R}^n$ or that there exists $u'_0 \in U_0$ such that $\hat{x}_1^k < f_0^k(\eta, u'_0)$ for all $k \in \{1, \ldots, m\}$, then we have $(\lambda_1, \ldots, \lambda_m) \neq (0, \ldots, 0)$. In the previous theorem, when $a = i$, we can replace the invertibility condition by the positivity condition as in Sect. 2.2.2.

The following result is a result of sufficient conditions.

Theorem 2.23. *Let $(\underline{\hat{x}}, \underline{\hat{u}}) \in \mathrm{Dom}^e_\eta((J_j)_{1 \leq j \leq m})$. We assume that there exist $(\lambda_1, \ldots, \lambda_m) \in \mathbb{R}^m$ and $(p_{t+1})_{t \in \mathbb{N}} \in (\mathbb{R}^{n*})^{\mathbb{N}}$ which satisfy the following conditions:*

(i) $(\lambda_1, \ldots, \lambda_m, p_1) \neq (0, \ldots, 0, 0)$.
(ii) For all $j \in \{1, \ldots, m\}$, $\lambda_j \geq 0$.
(iii) For all $t \in \mathbb{N}_*$, $p_t = \sum_{j=1}^{m} \lambda_j D_1\phi_{j,t}(\hat{x}_t, \hat{u}_t) + p_{t+1} \circ D_1 f_t(\hat{x}_t, \hat{u}_t)$.
(iv) For all $t \in \mathbb{N}$, for all $u_t \in U_t$, $\sum_{j=1}^{m} \lambda_j \phi_{j,t}(\hat{x}_t, \hat{u}_t) + \langle p_{t+1}, f_t(\hat{x}_t, \hat{u}_t) \rangle \geq \sum_{j=1}^{m} \lambda_j \phi_{j,t}(\hat{x}_t, u_t) + \langle p_{t+1}, f_t(\hat{x}_t, u_t) \rangle$.
(v) For all $t \in \mathbb{N}$, $X_t \times U_t$ is convex and the function

$(x_t, u_t) \mapsto \sum_{j=1}^{m} \lambda_j \phi_{j,t}(x_t, u_t) + \langle p_{t+1}, f_t(x_t, u_t)\rangle$ *is concave.*

(vi) For all $x_t \in X_t$, $\lim_{t\to+\infty} \langle p_{t+1}, x_t - \hat{x}_t\rangle = 0$.

Then $(\underline{\hat{x}}, \underline{\hat{u}})$ *is a solution of* $(\mathcal{V}_e^{n,w})$*, and moreover if* $\lambda_j > 0$ *for all* $j \in \{1, \ldots, m\}$ *then* $(\underline{\hat{x}}, \underline{\hat{u}})$ *is a solution of* $(\mathcal{V}_e^{n})$*.*

The proof of this result uses the well-known fact that is: if $(\underline{\hat{x}}, \underline{\hat{u}})$ maximizes the weighted functional $\sum_{j=1}^{m} \theta_j J_j(\underline{x}, \underline{u})$ where $\theta_j \geq 0$ for all $j \in \{1, \ldots, m\}$, then $(\underline{\hat{x}}, \underline{\hat{u}})$ is a weak Pareto optimum, i.e., a solution of $(\mathcal{V}_e^{n,w})$. The concavity condition permits to transform necessary conditions of optimality on the weighted functional into sufficient conditions of optimality. The assumption (vi) is called a sufficient condition of transversality at infinity.

Remark 2.8. Using the function

$$(x_t, p_{t+1}, \lambda_1, \ldots, \lambda_m) \mapsto \max_{u_t \in U_t} \left(\sum_{j=1}^{m} \lambda_j \phi_{j,t}(x_t, u_t) + \langle p_{t+1}, f_t(x_t, u_t)\rangle \right)$$

it is possible to state an additional theorem of sufficient conditions, [51].

To finish this section, we provide a strong Pontryagin principle in presence of constraints in the form $\mathcal{U}_t^1(x_t)$ as defined in (2.49). $\mathrm{Adm}^a_{\eta,c}$ is defined by replacing U_t by $\mathcal{U}_t^1(x_t)$ in Adm^a_η, $\mathrm{Dom}^a_{\eta,c}(J_j)$ is defined by replacing U_t by $\mathcal{U}_t^1(x_t)$ in $\mathrm{Dom}^a_\eta(J_j)$, and $\mathrm{Dom}^a_{\eta,c}((J_j)_{1\leq j\leq m}) := \bigcap_{1\leq j\leq m} \mathrm{Dom}^a_{\eta,c}(J_j)$. When $\ell \in \{n, o, w\}$ and $a \in \{e, i\}$, $(\mathcal{V}_a^{\ell,c})$ and $(\mathcal{V}_a^{\ell,w,c})$ are obtained by replacing Adm^a_η by $\mathrm{Adm}^a_{\eta,c}$ and $\mathrm{Dom}^a_\eta((J_j)_{1\leq j\leq m})$ by $\mathrm{Dom}^a_{\eta,c}((J_j)_{1\leq j\leq m})$ in $(\mathcal{V}_a^{\ell})$ and $(\mathcal{V}_a^{\ell,w})$. In the conditions of Michel, the sets $A_t(x_t, x_{t+1})$ and $B_t(x_t, x_{t+1})$ are defined as before in Theorem 2.20 in the previous section.

Theorem 2.24. *Let* $(\underline{\hat{x}}, \underline{\hat{u}})$ *be a solution of* $(\mathcal{V}_a^{n,w,c})$ *or of* $(\mathcal{V}_a^{o,w,c})$ *or of* $(\mathcal{V}_a^{w,w,c})$*. We assume that the following conditions are fulfilled:*

(a) For all $t \in \mathbb{N}$, X_t *is nonempty open and convex, and for all* $u_t \in U_t$*, for all* $j \in \{1, \ldots, m\}$, $\phi_{j,t}(., u_t)$, $f_t(., u_t)$ *and* $g_t(., u_t)$ *are of class* C^1 *on* X_t*.*
(b) For all $t \in \mathbb{N}$*, for all* $(x_t, x_{t+1}) \in X_t \times X_{t+1}$, $coA_t(x_t, x_{t+1}) \subset B_t(x_t, x_{t+1})$*.*
(c) For all $t \in \mathbb{N}$, $D_1 f_t(\hat{x}_t, \hat{u}_t)$ *is invertible.*
(d) For all $t \in \mathbb{N}$*, there exists* $u_t' \in U_t$ *such that* $f_t(\hat{x}_t, u_t') = f_t(\hat{x}_t, \hat{u}_t)$ *and* $g_t^h(\hat{x}_t, u_t') > 0$ *for all* $h \in \{1, \ldots, d^i\}$*.*

Then there exist $(\lambda_1, \ldots, \lambda_m) \in \mathbb{R}^m$, $(p_{t+1})_{t\in\mathbb{N}} \in (\mathbb{R}^{n*})^{\mathbb{N}}$ *and* $(q_t)_{t\in\mathbb{N}} \in \mathbb{R}^{d^i*}$ *which satisfy the following conditions:*

(i) $(\lambda_1, \ldots, \lambda_m, p_1) \neq (0, \ldots, 0, 0)$*.*
(ii) $\lambda_j \geq 0$ *for all* $j \in \{1, \ldots, m\}$*, and* $q_t \geq 0$ *for all* $t \in \mathbb{N}$*.*

(iii) $p_t = \sum_{j=1}^{m} \lambda_j D_1\phi_{j,t}(\hat{x}_t, \hat{u}_t) + p_{t+1}, \circ D_1 f_t(\hat{x}_t, \hat{u}_t) + q_t \circ D_1 g_t(\hat{x}_t, \hat{u}_t)$ *for all* $t \in \mathbb{N}$.

(iv) $\sum_{j=1}^{m} \lambda_j \phi_{j,t}(\hat{x}_t, \hat{u}_t) + \langle p_{t+1}, f_t(\hat{x}_t, \hat{u}_t)\rangle + \langle q_t, g_t(\hat{x}_t, \hat{u}_t)\rangle \geq$
$\sum_{j=1}^{m} \lambda_j \phi_{j,t}(\hat{x}_t, u_t) + \langle p_{t+1}, f_t(\hat{x}_t, u_t)\rangle + \langle q_t, g_t(\hat{x}_t, u_t)\rangle$ *for all* $u_t \in U_t$, *for all* $t \in \mathbb{N}$.

Moreover, if in addition we assume that $f_0(\{\eta\} \times U_0) - \hat{x}_1$ *is a neighborhood of 0 in* $\mathbb{R}^n$, *and if there exists* $u_0'' \in U_0$ *such that* $f_0(\eta, u_0'') - \hat{x}_1 = 0$ *and* $g_t^h(\eta, u_0'') > 0$ *for all* $h \in \{1, \ldots, d^i\}$, *then we have* $(\lambda_1, \ldots, \lambda_m) \neq (0, \ldots, 0)$.

The proof of this theorem also uses the reduction to finite horizon. In addition it uses a generalization of the parametrized static optimization Theorem 1.6 for single-objective problems to the multiobjective case. This generalization to weak Pareto optima can be found in Hayek [50].

Chapter 3
The Special Case of the Bounded Processes

3.1 Introduction

Infinite-horizon discrete-time optimal control problems in the set of bounded processes are examined. According to Chichilnisky [35, 36], the space of bounded sequences was first used in economics by Debreu [39]. It can also be found in [34], for example. From a mathematical point of view it allows to use analysis in Banach spaces. We establish necessary conditions of optimality for infinite-horizon discrete-time optimal control problems with state equation or state inequation, for bounded processes. It necessitates to manipulate the dual of ℓ_∞, to establish results on bounded solutions of difference equations of order one. We apply abstract optimization theorems in Banach spaces to obtain strong and weak Pontryagin principles in Sect. 3.2. In Sect. 3.3, for problems governed by inequations, we work in ordered Banach spaces and we treat the state inequation as an infinity of inequality constraints, by using abstract results of optimization theory in ordered Banach spaces in the spirit of the Karush–Kuhn–Tucker theorem. In Sect. 3.4, we provide links with unbounded problems and in Sect. 3.5 we give sufficient conditions of optimality. The mathematical tools used in this chapter belong to linear and nonlinear functional analysis: sequence spaces, Nemytskii's operators, duality in topological vector spaces, and ordered Banach spaces (see e.g. [32, 56, 57]).

3.2 The Bounded Case with (DE)

In this section, we establish necessary conditions of optimality for infinite-horizon discrete-time optimal control problems with state equation, for bounded processes. We apply an abstract optimization theorem in Banach spaces which is due to Ioffe and Tihomirov to obtain a strong Pontryagin principle [54] and a Karush–Kuhn–Tucker theorem in Banach spaces to obtain a weak principle. The techniques use

J. Blot and N. Hayek, *Infinite-Horizon Optimal Control in the Discrete-Time Framework*, SpringerBriefs in Optimization, DOI 10.1007/978-1-4614-9038-8_3,

nonlinear operators and functionals in Banach spaces of sequences. The adjoint variable is in the dual of $\ell^\infty(\mathbb{N}, \mathbb{R}^n)$. We show that its component which is in $\ell^1(\mathbb{N}, \mathbb{R}^n)$ satisfies the usual necessary conditions of optimality.

3.2.1 A Strong Pontryagin Principle

Let Ω be a nonempty open convex subset of $\mathbb{R}^n$ and U a nonempty compact subset of $\mathbb{R}^d$. For all $t \in \mathbb{N}$, let $X_t = \Omega$ and U_t be a nonempty subset of U. We consider the functions $\psi : \Omega \times U \to \mathbb{R}$ and $f : \Omega \times U \to \mathbb{R}^n$.

For every $\underline{x} \in \mathbb{R}^n$ define $C(\underline{x})$ as the closure of the set of all terms of the sequence $\underline{x}$. If $\underline{x} \in \ell^\infty(\mathbb{N}, \mathbb{R}^n)$, $C(\underline{x})$ is compact. For every $\underline{u} \in \mathbb{R}^d$ define $C(\underline{u})$ as the closure of the set of all terms of the sequence $\underline{u}$.

We set $\mathscr{X} = \{\underline{x} = (x_t)_t \in \ell^\infty(\mathbb{N}, \mathbb{R}^n),$ such that $C(\underline{x}) \subset \Omega\}$. $\mathscr{X}$ is thus the set of all bounded sequences which are in the interior of Ω. Note that $\mathscr{X}$ is a convex open subset of $\ell^\infty(\mathbb{N}, \mathbb{R}^n)$ since Ω is open and convex. We set $\mathscr{U} = \{\underline{u} = (u_t)_t \in \prod_{t=0}^{\infty} U_t\} \subset \ell^\infty(\mathbb{N}, \mathbb{R}^d)$.

The controlled dynamical system is

(DE) $\quad x_{t+1} = f(x_t, u_t)$

When we fix an initial state $\eta \in \Omega$, we denote by Adm_η^e the set of all processes $(\underline{x}, \underline{u}) \in \Omega^{\mathbb{N}} \times \mathscr{U}$ which satisfy (DE) at each time $t \in \mathbb{N}$ and such that $x_0 = \eta$.

Let $\beta \in (0, 1)$. We consider first the following problem:

$$
(\mathscr{B}_\mathbf{e}) \quad \begin{cases} \text{Maximize } J(\underline{x}, \underline{u}) = \sum_{t=0}^{+\infty} \beta^t \psi(x_t, u_t) \\ \text{when} \quad x_0 = \eta \\ \qquad\quad \forall t \in \mathbb{N},\ x_{t+1} = f(x_t, u_t) \\ \qquad\quad (\underline{x}, \underline{u}) \in \mathscr{X} \times \mathscr{U} \end{cases}
$$

which can be written

$(\mathscr{B}_\mathbf{e})$ Maximize $J(\underline{x}, \underline{u})$ when $(\underline{x}, \underline{u}) \in \mathrm{Adm}_\eta^e \cap (\mathscr{X} \times \mathscr{U})$.

The following theorem can be found in Blot and Hayek [24] where functions f_t, $t \in \mathbb{N}$, instead of f, are considered in (DE) together with additional hypotheses.

Theorem 3.1. *Let $(\underline{\hat{x}}, \underline{\hat{u}})$ be a solution of $(\mathscr{B}_\mathbf{e})$. Assume that*

(i) The mappings ψ and f are of class C^0 on $\Omega \times U$.
For all $u \in U$, the partial mappings $x \mapsto \psi(x, u)$ and $x \mapsto f(x, u)$ are of class C^1 on Ω.
The mappings $D_1\psi$ and $D_1 f$ are of class C^0 on $\Omega \times U$.

(ii) *For all $t \in \mathbb{N}$, for all $x_t \in \Omega$, for all $u'_t, u''_t \in U_t$, and for all $\alpha \in [0,1]$, there exists $u_t \in U_t$ such that*

$$\begin{aligned} \psi(x_t, u_t) &\geq \alpha\psi(x_t, u'_t) + (1-\alpha)\psi(x_t, u''_t) \\ f(x_t, u_t) &= \alpha f(x_t, u'_t) + (1-\alpha) f(x_t, u''_t) \end{aligned}$$

(iii) $\sup_{t\in\mathbb{N}} ||D_1 f(\hat{x}_t, \hat{u}_t)|| < 1$.

Then there exists $(p_t)_{t\in\mathbb{N}_*} \in \ell^1(\mathbb{N}_*, \mathbb{R}^{n*})$ such that

(a) For all $t \in \mathbb{N}_*$, $p_t = p_{t+1} \circ D_1 f(\hat{x}_t, \hat{u}_t) + \beta^t D_1\psi(\hat{x}_t, \hat{u}_t)$.
(b) For all $t \in \mathbb{N}$, for all $u_t \in U_t$,
$\beta^t\psi(\hat{x}_t, \hat{u}_t) + \langle p_{t+1}, f(\hat{x}_t, \hat{u}_t)\rangle \geq \beta^t\psi(\hat{x}_t, u_t) + \langle p_{t+1}, f(\hat{x}_t, u_t)\rangle$.
(c) $\lim_{t\to+\infty} p_t = 0$.

The Pontryagin Hamiltonian is $H_t : \mathbb{R}^n_+ \times \mathbb{R}^d \times \mathbb{R} \times \mathbb{R}^{n*} \to \mathbb{R}$, defined by $H_t(x, u, p, \lambda_0) = \lambda_0\beta^t\psi(x, u) + \langle p, f(x, u)\rangle$, the adjoint equation (a) is $p_t = D_1 H_t(\hat{x}_t, \hat{u}_t, p_{t+1}, 1)$, and the strong maximum principle (b) is $H_t(\hat{x}_t, \hat{u}_t, p_{t+1}, 1) \geq H_t(\hat{x}_t, u_t, p_{t+1}, 1)$. Since $(p_t)_{t\in\mathbb{N}_*} \in \ell^1(\mathbb{N}_*, \mathbb{R}^{n*})$ we necessarily have (c) $\lim_{t\to\infty} p_t = 0$ which is the transversality condition at infinity.

3.2.2 Proof of Theorem 3.1

The steps of the proof are the following:

First step: The optimal control problem can be written as the following abstract static optimization problem in a Banach space:

$$\begin{cases} \text{Maximize } J(\underline{x}, \underline{u}) \\ \quad \text{when} \quad F(\underline{x}, \underline{u}) = 0 \\ \qquad\qquad\quad (\underline{x}, \underline{u}) \in \mathcal{X} \times \mathcal{U} \end{cases}$$

This problem is in the form of Problem ($\mathcal{P}\mathcal{P}2$) of Sect. 1.4.5, where $\Gamma^0 = J$ and without inequality constraints. So we shall show that it satisfies all conditions of Theorem 1.8 of Sect. 1.4.5; this theorem is due to Ioffe and Tihomirov [54]. We show this through the following lemmas:

Lemma 3.1. *Under (i), for all $\underline{u} \in \mathcal{U}$, the functional $\underline{x} \mapsto J(\underline{x}, \underline{u})$ and the operator $\underline{x} \mapsto F(\underline{x}, \underline{u})$ defined by $F(\underline{x}, \underline{u}) := (f(x_t, u_t) - x_{t+1})_t$ are of class C^1 on $\mathcal{X}$ and moreover the following formulas hold, for all $(\underline{x}, \underline{u}) \in \mathcal{X} \times \mathcal{U}$, $\underline{\delta x} \in \ell^\infty(\mathbb{N}, \mathbb{R}^n)$:*

1. $D_1 J(\underline{x}, \underline{u})\underline{\delta x} = \sum_{t=0}^{+\infty} \beta^t D_1\psi(x_t, u_t)\delta x_t$.
2. $D_1 F(\underline{x}, \underline{u})\underline{\delta x} = (D_1 f(x_t, u_t)\delta x_t - \delta x_{t+1})_{t\in\mathbb{N}}$.

Proof. Under the hypotheses of the theorem, J is well defined and

$$D_1 J(\underline{x}, \underline{u})\delta\underline{x} = \sum_{t=0}^{+\infty} \beta^t D_1 \psi(x_t, u_t)\delta x_t.$$

(see [17]).

We set $F(\underline{x}, \underline{u}) = (f(x_t, u_t) - x_{t+1})_{t\in\mathbb{N}}$ for all $(\underline{x}, \underline{u}) \in \mathscr{X} \times \mathscr{U}$.

Let $(\underline{x}, \underline{u}) \in \mathscr{X} \times \mathscr{U}$. We have $C(\underline{x}) \subset \Omega$. Since f is continuous on $\Omega \times U$ and $C(\underline{x}) \times C(\underline{u})$ is compact, $f(C(\underline{x}) \times C(\underline{u}))$ is compact, therefore bounded so the set $\{f(x_t, u_t) \ : \ t \in \mathbb{N}\}$ is bounded since $\{f(x_t, u_t) \ : \ t \in \mathbb{N}\} \subset f(C(\underline{x}) \times C(\underline{u}))$. So there exists K such that $\forall t \in \mathbb{N}$, $||f(x_t, u_t) - x_{t+1}|| \le K + \sup_{t\in\mathbb{N}} ||x_t||$. So we have $F(\underline{x}, \underline{u}) \in \ell^\infty(\mathbb{N}, \mathbb{R}^n)$.

Let us show that $\underline{x} \mapsto F(\underline{x}, \underline{u})$ is of class C^0 on $\mathscr{X}$. Take $(\underline{x}^0, \underline{u}) \in \mathscr{X} \times \mathscr{U}$. Let $\epsilon > 0$ be given. Since $C(\underline{x}^0) \times C(\underline{u})$ is a compact of $\Omega \times U$, it follows from the Heine–Schwartz lemma [17] that there exists $\delta > 0$ such that for all $x_t \in \Omega$, for all $u_t \in U$, $||x_t - x_t^0|| = ||x_t - x_t^0|| + ||u_t - u_t|| < \delta$ implies $||f(x_t^0, u_t) - f(x_t, u_t)|| < \epsilon/2$. Let $\xi = \min\{\epsilon/2, \delta\}$ and let $\underline{x} \in \mathscr{X}$ be such that $||\underline{x} - \underline{x}^0||_\infty < \xi$. Then for all $t \in \mathbb{N}$, $||f(x_t^0, u_t) - f(x_t, u_t)|| < \epsilon/2$. So $||f(x_t, u_t) - x_{t+1} - f(x_t^0, u_t) + x_{t+1}^0|| \le ||f(x_t^0, u_t) - f(x_t, u_t)|| + ||x_{t+1} - x_{t+1}^0|| \le \epsilon$ which implies that $||F(\underline{x}, \underline{u}) - F(\underline{x}^0, \underline{u})||_\infty < \epsilon$. Let us now show that $\underline{x} \mapsto F(\underline{x}, \underline{u})$ is Fréchet differentiable on $\mathscr{X}$. Take $(\underline{x}^0, \underline{u}) \in \mathscr{X} \times \mathscr{U}$. Let $\epsilon > 0$ be given. Since $C(\underline{x}^0) \times C(\underline{u})$ is a compact of $\Omega \times U$, it follows from the Heine–Schwartz lemma that there exists $\delta' > 0$ such that for all $x_t \in \Omega$, for all $u_t \in U$, $||x_t - x_t^0|| = ||x_t - x_t^0|| + ||u_t - u_t|| < \delta'$ implies $||D_1 f(x_t^0, u_t) - D_1 f(x_t, u_t)|| < \epsilon$. Let $\underline{x} \in \mathscr{X}$ be such that $||\underline{x} - \underline{x}^0||_\infty < \delta'$. Then, for all $t \in \mathbb{N}$, $||f(x_t, u_t) - x_{t+1} + f(x_t^0, u_t) - x_{t+1}^0 + D_1 f(x_t^0, u_t)(x_t - x_t^0) - (x_{t+1} - x_{t+1}^0)|| \le (\sup_{y_t \in]x_t^0, x_t[} ||D_1 f(y_t, u_t) - D_1 f(x_t^0, u_t)||)||(x_t - x_t^0)|| \le \epsilon||(x_t - x_t^0)|| \le \epsilon||(\underline{x} - \underline{x}^0)||_\infty$. But this implies that

$$||F(\underline{x}, \underline{u}) - F(\underline{x}^0, \underline{u}) - (D_1 f(x_t^0, u_t)(x_t - x_t^0))_{t\in\mathbb{N}} - (x_{t+1} - x_{t+1}^0)_{t\in\mathbb{N}}||_\infty \le \epsilon||(\underline{x} - \underline{x}^0)||_\infty.$$

Thus $\underline{x} \mapsto F(\underline{x}, \underline{u})$ is Fréchet differentiable at $\underline{x}^0$ and

$$D_1 F(\underline{x}^0, \underline{u})\delta\underline{x} = (D_1 f(x_t^0, u_t)\delta x_t - \delta x_{t+1})_{t\in\mathbb{N}}.$$

To show the continuity of $\underline{x} \mapsto D_1 F(\underline{x}, \underline{u})$ at $\underline{x}^0$ let $\epsilon > 0$ be given and let $\underline{x} \in \mathscr{X}$ be such that $||\underline{x} - \underline{x}^0|| < \delta'$ where δ' is obtained above. Then $||D_1 F(\underline{x}, \underline{u}) - D_1 F(\underline{x}^0, \underline{u})||_\infty \le \sup_{t\in\mathbb{N}} ||D_1 f[x_t^0, u_t) - D_1 f(x_t, u_t)|| \le \sup_{t\in\mathbb{N}} \epsilon = \epsilon$. So $\underline{x} \mapsto D_1 F(\underline{x}, \underline{u})$ is continuous at $\underline{x}^0$. □

Lemma 3.2. *For all $\underline{x} \in \mathscr{X}$, for all $\underline{u}', \underline{u}'' \in \mathscr{U}$, and for all $\alpha \in [0, 1]$, there exists $\underline{u} \in \mathscr{U}$ such that*

$$J(\underline{x}, \underline{u}) \ge \alpha J(\underline{x}, \underline{u}') + (1 - \alpha) J(\underline{x}, \underline{u}'')$$
$$F(\underline{x}, \underline{u}) = \alpha F(\underline{x}, \underline{u}') + (1 - \alpha) F(\underline{x}, \underline{u}'')$$

Proof. Let $\underline{x} = (x_t)_t \in \mathcal{X}$, $\underline{u}' = (u'_t)_t \in \mathcal{U}$, $\underline{u}'' = (u''_t)_t \in \mathcal{U}$ and $\alpha \in [0, 1]$. Hypothesis (ii) of Theorem 3.1 implies for all $t \in \mathbb{N}$ the existence of $u_t \in U_t$ such that

$$\psi(x_t, u_t) \geq \alpha\psi(x_t, u'_t) + (1-\alpha)\psi(x_t, u''_t)$$
$$f(x_t, u_t) = \alpha f(x_t, u'_t) + (1-\alpha) f(x_t, u''_t).$$

Therefore we obtain

$$\sum_{t=0}^{+\infty} \beta^t \psi(x_t, u_t) \geq \alpha \sum_{t=0}^{+\infty} \beta^t \psi(x_t, u'_t) + (1-\alpha) \sum_{t=0}^{+\infty} \beta^t \psi(x_t, u''_t)$$
$$(f(x_t, u_t) - x_{t+1})_t = \alpha(f(x_t, u'_t) - x_{t+1})_t + (1-\alpha)(f(x_t, u''_t) - x_{t+1})_t.$$

Set $\underline{u} = (u_t)_t$, so $\underline{u} \in \mathcal{U}$ and satisfies the required relations. □

Lemma 3.3. *Under hypotheses (i) and (iii), we have* $\mathrm{Im}(D_1F(\hat{\underline{x}}, \hat{\underline{u}})) = \ell^\infty(\mathbb{N}, \mathbb{R}^n)$.

Proof. Since $D_1F(\underline{x}, \underline{u})\delta\underline{x} = (D_1 f(x_t, u_t)\delta x_t - \delta x_{t+1})_{t\in\mathbb{N}}$, $\delta x_0 = 0$, the problem is a problem of bounded solutions of first-order linear difference equations.

So we study bounded solutions of the following linear difference equations:

$$h_{t+1} - M_t h_t = b_t, \qquad h_0 = 0,$$

Let $(M_t)_{t\in\mathbb{N}} \in \ell^\infty(\mathbb{N}, (\mathbb{R}^n, \mathbb{R}^n))$. Assume that $\sup_{t\in\mathbb{N}_*} ||M_t|| < 1$. Then for all $(b_t)_{t\in\mathbb{N}} \in \ell^\infty(\mathbb{N}, \mathbb{R}^n)$ there exists a unique $(h_t)_{t\in\mathbb{N}} \in \ell^\infty(\mathbb{N}, \mathbb{R}^n)$ such that for all $t \in \mathbb{N}$,

$$h_{t+1} - M_t h_t = b_t$$

where $h_0 = 0$.

Consider the operator $\mathcal{T} : \ell^\infty(\mathbb{N}_*, \mathbb{R}^n) \longrightarrow \ell^\infty(\mathbb{N}_*, \mathbb{R}^n)$ such that for all $\underline{h} \in \ell^\infty(\mathbb{N}_*, \mathbb{R}^n)$,

$$\mathcal{T}(\underline{h}) = (h_t - M_{t-1}h_{t-1})_{t\in\mathbb{N}_*}$$

$\mathcal{T} = I + T$ where

$$I = \text{identity operator of } \ell^\infty(\mathbb{N}_*, \mathbb{R}^n)$$
$$T(\underline{h}) := (0, -M_1h_1, -M_2h_2, \ldots, -M_th_t, \ldots)$$

Recall that the norm of a linear operator S between normed spaces is defined by $||S||_{\mathcal{L}} = \sup_{||\underline{z}||\leq 1} ||S(\underline{z})||$. So $||T(\underline{h})||_{\ell^\infty} = \sup_{t\in\mathbb{N}_*} || - M_t h_t|| \leq (\sup_{t\in\mathbb{N}_*} ||M_t||)||\underline{h}||_\infty$. So $||T||_{\mathcal{L}} \leq \sup_{t\in\mathbb{N}_*} ||M_t|| < 1$. Since $\mathcal{T} = I + T$ and $||T||_{\mathcal{L}} < 1$, $\mathcal{T}$ is invertible so it is surjective.

Set $M_t = D_1 f(\hat{x}_t, \hat{u}_t)$. Then under (iv) we have $\sup_{t\in\mathbb{N}_*} ||D_1 f(\hat{x}_t, \hat{u}_t)|| < 1$. So $\mathcal{T}$ is surjective that is $\mathrm{Im}(D_1F(\hat{\underline{x}}, \hat{\underline{u}})) = \ell^\infty(\mathbb{N}, \mathbb{R}^n)$. □

Second step: In this step we can apply Theorem 1.8 and obtain the existence of $\lambda_0 \in \mathbb{R}$, $\Lambda \in \ell^\infty(\mathbb{N}, \mathbb{R}^n)^*$, not all zero, $\lambda_0 \geq 0$, such that

(AE) $\lambda_0 D_1 J(\underline{\hat{x}}, \underline{\hat{u}}) + \Lambda \circ D_1 F(\underline{\hat{x}}, \underline{\hat{u}}) = 0$
and
(PMP) for all $\underline{u} \in \mathscr{U}$, $(\lambda_0 J + \langle \Lambda, F \rangle)(\underline{\hat{x}}, \underline{\hat{u}}) \geq (\lambda_0 J + \langle \Lambda, F \rangle)(\underline{\hat{x}}, \underline{u})$

(AE) denotes the adjoint equation of this problem and (PMP) the Pontryagin maximum principle.

Using (AE), (PMP), and the formulas of operators given in Lemma 3.1, we obtain the following relations:

$$\left. \begin{array}{c} \text{For all } \underline{\delta x} \in \ell^\infty(\mathbb{N}, \mathbb{R}^n) \text{ such that } \delta x_0 = 0, \\ \lambda_0 \sum_{t=0}^{+\infty} \beta^t D_1 \psi(\hat{x}_t, \hat{u}_t) \delta x_t + \langle \Lambda, (D_1 f(\hat{x}_t, \hat{u}_t) \delta x_t - \delta x_{t+1})_{t \in \mathbb{N}} \rangle = 0 \end{array} \right\} \quad (3.1)$$

and

$$\left. \begin{array}{c} \text{For all } u_t \in U_t, \\ \lambda_0 \sum_{t=0}^{+\infty} \beta^t \psi(\hat{x}_t, \hat{u}_t) + \langle \Lambda, (f(\hat{x}_t, \hat{u}_t) - \hat{x}_{t+1})_{t \in \mathbb{N}} \rangle \geq \\ \lambda_0 \sum_{t=0}^{+\infty} \beta^t \psi(\hat{x}_t, u_t) + \langle \Lambda, (f(\hat{x}_t, u_t) - \hat{x}_{t+1})_{t \in \mathbb{N}} \rangle. \end{array} \right\} \quad (3.2)$$

Using Proposition A.3 we know that there exist $\underline{p} \in \ell^1(\mathbb{N}, \mathbb{R}^{n*})$ and $\theta \in \ell^1_d(\mathbb{N}, \mathbb{R}^n)$ such that $\Lambda = \underline{p} + \theta$. So (3.1) and (3.2) can be written as

$$\left. \begin{array}{c} \text{For all } \underline{\delta x} \in \ell^\infty(\mathbb{N}, \mathbb{R}^n), \text{ with } \delta x_0 = 0, \\ \lambda_0 \sum_{t=0}^{+\infty} \beta^t D_1 \psi(\hat{x}_t, \hat{u}_t) \delta x_t + \sum_{t=0}^{+\infty} \langle p_{t+1}, D_1 f(\hat{x}_t, \hat{u}_t) \delta x_t \rangle \\ - \sum_{t=0}^{+\infty} \langle p_{t+1}, \delta x_{t+1} \rangle + \langle \theta, (D_1 f(\hat{x}_t, \hat{u}_t) \delta x_t - \delta x_{t+1})_{t \in \mathbb{N}} \rangle = 0 \end{array} \right\} \quad (3.3)$$

and

$$\left. \begin{array}{c} \text{For all } u_t \in U_t, \\ \lambda_0 \sum_{t=0}^{+\infty} \beta^t (\psi(\hat{x}_t, \hat{u}_t) - \psi(\hat{x}_t, u_t)) + \langle \underline{p}, (f(\hat{x}_t, \hat{u}_t) - f(\hat{x}_t, u_t))_{t \in \mathbb{N}} \rangle + \\ \langle \theta, (f(\hat{x}_t, \hat{u}_t) - f(\hat{x}_t, u_t))_{t \in \mathbb{N}} \rangle \geq 0. \end{array} \right\} \quad (3.4)$$

Thus (3.3) becomes

$$\left.\begin{array}{l} \text{For all } \underline{\delta x} \in \ell^\infty(\mathbb{N}, \mathbb{R}^n) \text{ such that } \delta x_0 = 0, \\ \sum_{t=1}^{+\infty} \langle \lambda_0 \beta^t D_1\psi(\hat{x}_t, \hat{u}_t) + p_{t+1} \circ D_1 f(\hat{x}_t, \hat{u}_t) - p_t, \delta x_t \rangle \\ \qquad = -\langle \theta, (D_1 f(\hat{x}_t, \hat{u}_t)\delta x_t - \delta x_{t+1})_{t\in\mathbb{N}} \rangle. \end{array}\right\} \tag{3.5}$$

Now we arbitrarily fix $t \in \mathbb{N}_*$, we fix $\delta x_s = 0$ when $s \in \mathbb{N} \setminus \{t\}$, and δx_t varies in $\mathbb{R}^n$. Then we have $D_1 f(\hat{x}_s, \hat{u}_s)\delta x_s - \delta x_{s+1} = 0$ when $s \notin \{t-1, t\}$; hence $(D_1 f(\hat{x}_s, \hat{u}_s)\delta x_s - \delta x_{s+1})_s \in c_0(\mathbb{N}, \mathbb{R}^n)$. Thus $\langle \theta, (D_1 f(\hat{x}_s, \hat{u}_s)\delta x_s - \delta x_{s+1})_s \rangle = \langle c, \lim_{s\to\infty} (D_1 f(\hat{x}_s, \hat{u}_s)\delta x_s - \delta x_{s+1}) \rangle = \langle c, 0 \rangle = 0$ (see Definition A.2 in Appendix A).

So

$$\begin{aligned} 0 &= \sum_{s=1}^{+\infty} \langle \lambda_0 \beta^s D_1\psi(\hat{x}_s, \hat{u}_s) + p_{s+1} \circ D_1 f(\hat{x}_s, \hat{u}_s) - p_s, \delta x_s \rangle \\ &= \langle \lambda_0 \beta^t D_1\psi(\hat{x}_t, \hat{u}_t) + p_{t+1} \circ D_1 f(\hat{x}_t, \hat{u}_t) - p_t, \delta x_t \rangle. \end{aligned}$$

Therefore for all $t \in \mathbb{N}_*$ and for all $\delta x_t \in \mathbb{R}^n$ we have

$$\langle \lambda_0 \beta^t D_1\psi(\hat{x}_t, \hat{u}_t) + p_{t+1} \circ D_1 f(\hat{x}_t, \hat{u}_t) - p_t, \delta x_t \rangle = 0$$

And so we have proven the following relation:

$$\text{For all } t \in \mathbb{N}_*, \ p_t = p_{t+1} \circ D_1 f(\hat{x}_t, \hat{u}_t) + \lambda_0 \beta^t D_1\psi(\hat{x}_t, \hat{u}_t).$$

Now we arbitrarily fix $t \in \mathbb{N}$, we take $u_s = \hat{u}_s$ when $s \in \mathbb{N} \setminus \{t\}$, and u_t varies in U_t. Thus $(f(\hat{x}_t, \hat{u}_t) - f(\hat{x}_t, u_t))_{t\in\mathbb{N}}$ is of finite support and so $\langle \theta, (f(\hat{x}_t, \hat{u}_t) - f(\hat{x}_t, u_t))_{t\in\mathbb{N}} \rangle = 0$. Therefore

$$\begin{aligned} 0 &\leq \lambda_0 \sum_{s=0}^{+\infty} \beta^s (\psi(\hat{x}_s, \hat{u}_s) - \psi(\hat{x}_s, u_s)) \\ &+ \langle \underline{p}, (f(\hat{x}_s, \hat{u}_s) - f(\hat{x}_s, u_s))_{s\in\mathbb{N}} \rangle + \langle \theta, (f(\hat{x}_s, \hat{u}_s) - f(\hat{x}_s, u_s))_{s\in\mathbb{N}} \rangle \\ &= \lambda_0 \beta^t (\psi(\hat{x}_t, \hat{u}_t) - \psi(\hat{x}_t, u_t)) + \langle p_{t+1}, f(\hat{x}_t, \hat{u}_t) - f(\hat{x}_t, u_t) \rangle. \end{aligned}$$

And so we have proven the following relation:

$$\text{For all } u_t \in U_t, \ \lambda_0 \beta^t \psi(\hat{x}_t, \hat{u}_t) + \langle p_{t+1}, f(\hat{x}_t, \hat{u}_t) \rangle \geq \lambda_0 \beta^t \psi(\hat{x}_t, u_t) + \langle p_{t+1}, f(\hat{x}_t, u_t) \rangle.$$

Third step:

Lemma 3.4. $\lambda_0 \neq 0$.

Proof. Recall we obtained the existence of $\lambda_0 \in \mathbb{R}$, $\Lambda \in \ell^\infty(\mathbb{N}, \mathbb{R}^n)^*$, not all zero, $\lambda_0 \geq 0$, such that $\lambda_0 D_1 J(\underline{\hat{x}}, \underline{\hat{u}}) + \Lambda \circ D_1 F(\underline{\hat{x}}, \underline{\hat{u}}) = 0$. If $\lambda_0 = 0$, then $\Lambda = 0$ since $\mathrm{Im}(D_1 F(\underline{\hat{x}}, \underline{\hat{u}})) = \ell^\infty(\mathbb{N}, \mathbb{R}^n)$. Hence $\lambda_0 \neq 0$. We can set it equal to one. □

Therefore conclusions (a) and (b) are satisfied.

Conclusion (c) is a straightforward consequence of the belonging of $(p_t)_{t\in\mathbb{N}_*}$ to $\ell^1(\mathbb{N}_*, \mathbb{R}^{n*})$.

Lemma 3.5. $\theta = 0$

Proof. Using conclusion (a) and (3.5) we obtain $\langle \theta, (D_1 f(\hat{x}_t, \hat{u}_t)\delta x_t - \delta x_{t+1})_{t\in\mathbb{N}} \rangle = 0$, for all $\underline{\delta x} \in \ell^\infty(\mathbb{N}, \mathbb{R}^n)$.

Using $\mathrm{Im} D_1 F(\underline{\hat{x}}, \underline{\hat{u}}) = \ell^\infty(\mathbb{N}, \mathbb{R}^n)$ we have $\langle \theta, \underline{h} \rangle = 0$, for all $\underline{h} \in \ell^\infty(\mathbb{N}, \mathbb{R}^n)$. Thus $\theta = 0$. □

3.2.3 A Weak Pontryagin Principle

In this section we set $X_t = \mathbb{R}^n$, $U_t = \mathbb{R}^d$. So $\mathscr{X} \times \mathscr{U} = \ell^\infty(\mathbb{N}, \mathbb{R}^n) \times \ell^\infty(\mathbb{N}, \mathbb{R}^d)$ and we consider problem $(\mathscr{B}_e)$ of the previous section.

Theorem 3.2. *Let $(\underline{\hat{x}}, \underline{\hat{u}}) \in \ell^\infty(\mathbb{N}, \mathbb{R}^n) \times \ell^\infty(\mathbb{N}, \mathbb{R}^d)$ be a solution of problem $(\mathscr{B}_e)$. Assume that*

(i) The mappings ψ and f are of class C^1 on $\mathbb{R}^n \times \mathbb{R}^d$.
(ii) $\sup_{t\in\mathbb{N}} \|D_1 f(\hat{x}_t, \hat{u}_t)\| < 1$.

Then there exists $(p_t)_{t\in\mathbb{N}_*} \in \ell^1(\mathbb{N}_*, \mathbb{R}^{n*})$ such that

(a) For all $t \in \mathbb{N}_*$, $p_t = p_{t+1} \circ D_1 f(\hat{x}_t, \hat{u}_t) + \beta^t D_1\psi(\hat{x}_t, \hat{u}_t)$.
(b) For all $t \in \mathbb{N}$, $p_{t+1} \circ D_2 f(\hat{x}_t, \hat{u}_t) + \beta^t D_2\psi(\hat{x}_t, \hat{u}_t) = 0$.
(c) $\lim_{t\to+\infty} p_t = 0$.

The Pontryagin Hamiltonian is $H_t : \mathbb{R}^n_+ \times \mathbb{R}^d \times \mathbb{R} \times \mathbb{R}^{n*} \to \mathbb{R}$, defined by $H_t(x, u, p, \lambda_0) = \lambda_0 \beta^t \psi(x, u) + \langle p, f(x, u) \rangle$, the adjoint equation (a) is $p_t = D_1 H_t(\hat{x}_t, \hat{u}_t, p_{t+1}, 1)$, and the weak maximum principle (b) is $D_2 H_t(\hat{x}_t, \hat{u}_t, p_{t+1}, 1) = 0$. Since $(p_t)_{t\in\mathbb{N}_*} \in \ell^1(\mathbb{N}_*, \mathbb{R}^{n*})$ we necessarily have (c) $\lim_{t\to\infty} p_t = 0$ which is the transversality condition at infinity.

3.2.4 Proof of Theorem 3.2

The steps of the proof are the following:

First step: The optimal control problem can be written as the following abstract static optimization problem in a Banach space:

$$\begin{cases} \text{Maximize } J(\xi) \\ \text{when} \qquad F(\xi) = 0 \\ \qquad\qquad \xi \in \Xi \end{cases}$$

We shall show that it satisfies all conditions of Theorem B.10 (Appendix B) called Lagrange principle in [1]. We set $\Xi := \ell^\infty(\mathbb{N}, \mathbb{R}^n) \times \ell^\infty(\mathbb{N}, \mathbb{R}^d)$; it is a Banach space as a product of two Banach spaces. We set $Y := \ell^\infty(\mathbb{N}, \mathbb{R}^n)$. We set $F(\xi) = F(\underline{x}, \underline{u}) = (f(x_t, u_t) - x_{t+1})_t$ for all $(\underline{x}, \underline{u}) \in \ell^\infty(\mathbb{N}, \mathbb{R}^n) \times \ell^\infty(\mathbb{N}, \mathbb{R}^d)$.

Lemma 3.6. *Under (i), the functional J, the Nemytskii's operator N_f : $\ell^\infty(\mathbb{N}, \mathbb{R}^n) \times \ell^\infty(\mathbb{N}, \mathbb{R}^d) \to \ell^\infty(\mathbb{N}, \mathbb{R}^n)$, defined by $N_f(\underline{x}, \underline{u}) := (f(x_t, u_t))_t$, and the operator $F : \ell^\infty(\mathbb{N}, \mathbb{R}^n) \times \ell^\infty(\mathbb{N}, \mathbb{R}^d) \to \ell^\infty(\mathbb{N}, \mathbb{R}^n)$, defined by $F(\underline{x}, \underline{u}) := (f(x_t, u_t) - x_{t+1})_t$, are of class C^1 on $\ell^\infty(\mathbb{N}, \mathbb{R}^n) \times \ell^\infty(\mathbb{N}, \mathbb{R}^d)$, and moreover the following formulas hold, for all $(\underline{x}, \underline{u})$, $(\underline{\delta x}, \underline{\delta u}) \in \ell^\infty(\mathbb{N}, \mathbb{R}^n) \times \ell^\infty(\mathbb{N}, \mathbb{R}^d)$:*

1. $DJ(\underline{x}, \underline{u}).(\underline{\delta x}, \underline{\delta u}) = \sum_{t=0}^{\infty} \beta^t D_1\psi(x_t, u_t)\delta x_t + \sum_{t=0}^{\infty} \beta^t D_2\psi(x_t, u_t)\delta u_t$.
2. $DN_f(\underline{x}, \underline{u})(\underline{\delta x}, \underline{\delta u}) = (D_1 f(x_t, u_t).\delta x_t)_t + (D_2 f(x_t, u_t).\delta u_t)_t$.
3. $DF(\underline{x}, \underline{u})(\underline{\delta x}, \underline{\delta u}) = (D_1 f(x_t, u_t)\delta x_t - \delta x_{t+1})_t + (D_2 f(x_t, u_t).\delta u_t)_t$.

Proof. Using Theorem AI.2 in Appendix 1, p. 24, in [17], we obtain that N_f and the Nemytskii's operator $N_\psi : \ell^\infty(\mathbb{N}, \mathbb{R}^n) \times \ell^\infty(\mathbb{N}, \mathbb{R}^d) \to \ell^\infty(\mathbb{N}, \mathbb{R}^n)$, defined by $N_\psi(\underline{x}, \underline{u}) := (\psi(x_t, u_t))_t$, are of class C^1 and, for all $(\underline{x}, \underline{u})$, $(\underline{\delta x}, \underline{\delta u}) \in \ell^\infty(\mathbb{N}, \mathbb{R}^n) \times \ell^\infty(\mathbb{N}, \mathbb{R}^d)$, we have

$$\begin{aligned} DN_\psi(\underline{x}, \underline{u})(\underline{\delta x}, \underline{\delta u}) &= (D\psi(x_t, u_t)(\delta x_t, \delta u_t))_t \\ &= (D_1\psi(x_t, u_t)\delta x_t)_t + (D_2\psi(x_t, u_t)\delta u_t)_t, \end{aligned}$$

and

$$\begin{aligned} DN_f(\underline{x}, \underline{u})(\underline{\delta x}, \underline{\delta u}) &= (Df(x_t, u_t)(\delta x_t, \delta u_t))_t \\ &= (D_1 f(x_t, u_t)\delta x_t)_t + (D_2 f(x_t, u_t)\delta u_t)_t. \end{aligned}$$

The functional $S : \ell^\infty(\mathbb{N}, \mathbb{R}) \to \mathbb{R}$, defined by $S(\underline{r}) := \sum_{t=0}^{\infty} \beta^t r_t$, is linear. Since $|S(\underline{r})| \leq (\sum_{t=0}^{\infty} \beta^t).\|\underline{r}\|_{\ell^\infty} = \frac{1}{1-\beta}\|\underline{r}\|_{\ell^\infty}$, S is continuous. Therefore S is of class C^1, and we have, for all $\underline{r}$, $\underline{\delta r} \in \ell^\infty(\mathbb{N}, \mathbb{R})$; we have

$$DS(\underline{r})(\underline{\delta r}) = S(\underline{\delta r}) = \sum_{t=0}^{\infty} \beta^t.\delta r_t.$$

And so $J = S \circ N_\psi$ is of class C^1 as a composition of two C^1-mappings. Using the chain rule we obtain, for all $(\underline{x}, \underline{u})$, $(\underline{\delta x}, \underline{\delta u}) \in \ell^\infty(\mathbb{N}, \mathbb{R}^n) \times \ell^\infty(\mathbb{N}, \mathbb{R}^d)$, we have

$$\begin{aligned}
DJ(\underline{x},\underline{u})(\underline{\delta x},\underline{\delta u}) &= DS(N_\psi(\underline{x},\underline{u})) \circ DN_\psi(\underline{x},\underline{u})(\underline{\delta x},\underline{\delta u}) \\
&= S(DN_\psi(\underline{x},\underline{u})(\underline{\delta x},\underline{\delta u})) = \sum_{t=0}^{\infty} \beta^t D_1\psi(x_t,u_t)\delta x_t + \sum_{t=0}^{\infty} \beta^t D_2\psi(x_t,u_t)\delta u_t
\end{aligned}$$

And so, the result is proven for f and J. We introduce the operator $A : \ell^\infty(\mathbb{N},\mathbb{R}^n)\times \ell^\infty(\mathbb{N},\mathbb{R}^d) \to \ell^\infty(\mathbb{N},\mathbb{R}^n)$, defined by $A(\underline{x},\underline{u}) := (-x_{t+1})_t$.

A is linear and from $\|A(\underline{x},\underline{u})\|_{\ell^\infty} \le \|\underline{x}\|_{\ell^\infty} \le \|\underline{x}\|_{\ell^\infty} + \|\underline{u}\|_{\ell^\infty}$, we obtain that A is continuous. Consequently A is of class C^1 and, for all $(\underline{x},\underline{u})$, $(\underline{\delta x},\underline{\delta u}) \in \ell^\infty(\mathbb{N},\mathbb{R}^n)\times \ell^\infty(\mathbb{N},\mathbb{R}^d)$, we have

$$DA(\underline{x},\underline{u})(\underline{\delta x},\underline{\delta u}) = A(\underline{\delta x},\underline{\delta u}) = (-\delta x_{t+1})_t.$$

Note that $F = N_f + A$. And so F is of class C^1 as a sum of two C^1-mappings and, for all $(\underline{x},\underline{u})$, $(\underline{\delta x},\underline{\delta u}) \in \ell^\infty(\mathbb{N},\mathbb{R}^n)\times \ell^\infty(\mathbb{N},\mathbb{R}^d)$, we obtain

$$\begin{aligned}
DF(\underline{x},\underline{u})(\underline{\delta x},\underline{\delta u}) &= DN_f(\underline{x},\underline{u})(\underline{\delta x},\underline{\delta u}) + A(\underline{\delta x},\underline{\delta u}) \\
&= (D_1 f(x_t,u_t)\delta x_t)_t + (D_2 f(x_t,u_t)\delta u_t)_t + (-\delta x_{t+1})_t \\
&= (D_1 f(x_t,u_t)\delta x_t)_t - (\delta x_{t+1})_t + (D_2 f(x_t,u_t)\delta u_t)_t.
\end{aligned}$$ □

Lemma 3.7. *Under hypotheses (i) and (ii), we have* $\mathrm{Im}(DF(\underline{\hat{x}},\underline{\hat{u}})) = \ell^\infty(\mathbb{N},\mathbb{R}^n)$.

Proof. Under (i), $DF(\underline{\hat{x}},\underline{\hat{u}})(\underline{\delta x},\underline{\delta u}) = D_1F(\underline{x},\underline{u})\underline{\delta x} + D_2F(\underline{x},\underline{u})\underline{\delta u}$, and under (ii) and lemma 3.3, we have $\mathrm{Im}(D_1F(\underline{\hat{x}},\underline{\hat{u}})) = \ell^\infty(\mathbb{N},\mathbb{R}^n)$ so $\mathrm{Im}(DF(\underline{\hat{x}},\underline{\hat{u}})) = \ell^\infty(\mathbb{N},\mathbb{R}^n)$. □

Second step: In this step we can apply Theorem B.10 (Appendix B) and obtain the existence of $\lambda_0 \in \mathbb{R}$, $\Lambda \in \ell^\infty(\mathbb{N},\mathbb{R}^n)^*$, not all zero, $\lambda_0 \ge 0$, such that

$$\lambda_0 DJ(\hat{\underline{\xi}}) + \Lambda \circ DF(\hat{\underline{\xi}}) = 0$$

Using the formulas of operators given in Lemma 3.6, we obtain the following relation, for all $(\underline{x},\underline{u}) \in \ell^\infty(\mathbb{N},\mathbb{R}^n)\times \ell^\infty(\mathbb{N},\mathbb{R}^d)$:

$$\left.\begin{aligned}
0 = \lambda_0 . \sum_{t=0}^{\infty} \beta^t D\psi(\hat{x}_t,\hat{u}_t)(x_t-\hat{x}_t,u_t-\hat{u}_t) \\
+\langle P, (Df(\hat{x}_t,\hat{u}_t)(x_t-\hat{x}_t,u_t-\hat{u}_t) - (x_{t+1}-\hat{x}_{t+1}))_t\rangle
\end{aligned}\right\} \tag{3.6}$$

Using Proposition A.3 (Appendix A) we know that there exist $\underline{p} \in \ell^1(\mathbb{N},\mathbb{R}^{n*})$ and $\theta \in \ell^1_d(\mathbb{N},\mathbb{R}^n)$ such that $\Lambda = \underline{p} + \theta$, so to abridge the writing, we introduce, for all $t \in \mathbb{N}$,

$$d_t := Df(\hat{x}_t,\hat{u}_t)(x_t-\hat{x}_t,u_t-\hat{u}_t) - (x_{t+1}-\hat{x}_{t+1}). \tag{3.7}$$

We arbitrarily fix $t \in \mathbb{N}_*$, we fix $\underline{u} = \underline{\hat{u}}$, and we fix $x_s = \hat{x}_s$ when $s \in \mathbb{N} \setminus \{t\}$, and x_t varies in $\mathbb{R}^n$. Then we have $d_s = 0$ when $s \notin \{t-1, t\}$, $d_{t-1} = -(x_t - \hat{x}_t)$ and $d_t = D_1 f(\hat{x}_t, \hat{u}_t)(x_t - \hat{x}_t)$. And so we have $\underline{d} \in c_0(\mathbb{N}, \mathbb{R}^n)$ and consequently $\theta(\underline{d}) = \langle c, \lim_{s\to\infty} d_s\rangle = \langle c, 0\rangle = 0$ (see Definition A.2 in Appendix A) and $\langle \underline{p}, \underline{d}\rangle = \langle p_{t-1}, -(x_t - \hat{x}_t)\rangle + \langle p_t, D_1 f(\hat{x}_t, \hat{u}_t)(x_t - \hat{x}_t)\rangle = \langle -p_{t-1} + p_t \circ D_1 f(\hat{x}_t, \hat{u}_t), x_t - \hat{x}_t\rangle$. And so we have proven the following relation, for all $x_t \in \mathbb{R}^n$:

$$\langle \Lambda, \underline{d}\rangle = \langle -p_{t-1} + p_t \circ D_1 f(\hat{x}_t, \hat{u}_t), x_t - \hat{x}_t\rangle.$$

Moreover we have $\lambda_0 . \sum_{t=0}^{\infty} \beta^t D\psi(\hat{x}_t, \hat{u}_t).(x_t - \hat{x}_t, u_t - \hat{u}_t) = \lambda_0.\beta^t.D_1\psi(\hat{x}_t, \hat{u}_t).(x_t - \hat{x}_t)$. And then we obtain the following relation, for all $t \in \mathbb{N}_*$, for all $x_t \in \mathbb{R}^n$:

$$\langle \lambda_0.\beta^t D_1\psi(\hat{x}_t, \hat{u}_t) - p_{t-1} + p_t \circ D_1 f(\hat{x}_t, \hat{u}_t), x_t - \hat{x}_t\rangle = 0,$$

which gives

$$-p_{t-1} + p_t \circ D_1 f(\hat{x}_t, \hat{u}_t) + \lambda_0.\beta^t D_1\psi(\hat{x}_t, \hat{u}_t) = 0. \tag{3.8}$$

Now we take $\underline{x} = \underline{\hat{x}}$, we arbitrarily fix $t \in \mathbb{N}$, and we take $u_s = \hat{u}_s$ when $s \in \mathbb{N} \setminus \{t\}$, and u_t varies in $\mathbb{R}^d$. Then

$$\lambda_0 . \sum_{t=0}^{\infty} \beta^t D\psi(\hat{x}_t, \hat{u}_t)(x_t - \hat{x}_t, u_t - \hat{u}_t) = \lambda_0.\beta^t.D_2\psi(\hat{x}_t, \hat{u}_t)(u_t - \hat{u}_t).$$

When $s \neq t$ we have $d_s = Df(\hat{x}_s, \hat{u}_s)(0,0) - 0 = 0$ and $d_t = D_2 f(\hat{x}_t, \hat{u}_t)(u_t - \hat{u}_t)$. And so we have $\underline{d} \in c_0(\mathbb{N}, \mathbb{R}^n)$ which implies $\theta(\underline{d}) = 0$, and $\langle \underline{q}, \underline{d}\rangle = \langle p_t, d_t\rangle = \langle p_t, D_2 f(\hat{x}_t, \hat{u}_t)(u_t - \hat{u}_t)\rangle = \langle p_t \circ D_2 f(\hat{x}_t, \hat{u}_t), u_t - \hat{u}_t\rangle$. Then we deduce the following relation, for all $u_t \in \mathbb{R}^d$:

$$\langle \lambda_0.\beta^t D_2\psi(\hat{x}_t, \hat{u}_t) + p_t \circ D_2 f(\hat{x}_t, \hat{u}_t), u_t - \hat{u}_t\rangle = 0,$$

which gives

$$\lambda_0.\beta^t D_2\psi(\hat{x}_t, \hat{u}_t) + p_t \circ D_2 f(\hat{x}_t, \hat{u}_t) = 0. \tag{3.9}$$

Lemma 3.8. $\lambda_0 \neq 0$.

Proof. Since $\mathrm{Im}(DF(\underline{\hat{x}}, \underline{\hat{u}})) = \ell^\infty(\mathbb{N}, \mathbb{R}^n)$ we have $\lambda_0 \neq 0$ (see Theorem B.10). We can set it equal to one. □

Therefore conclusions (a) and (b) are satisfied.

Conclusion (c) is a straightforward consequence of the belonging of $(p_t)_{t\in\mathbb{N}_*}$ to $\ell^1(\mathbb{N}_*, \mathbb{R}^{n*})$.

3.3 The Bounded Case with (DI)

3.3.1 A Weak Pontryagin Principle

In this section we establish necessary conditions of optimality for infinite-horizon discrete-time optimal control problems with state inequation, for bounded processes. A major difference between our necessary conditions theorem and results established in the previous section is that we do not need the assumption: $\sup_{t\in\mathbb{N}} \|D_1 f(\hat{x}_t, \hat{u}_t)\| < 1$ which is essential in Theorems 3.1 and 3.2. In the previous section the system is governed by a difference equation, and the following theorem concerns a system governed by a difference inequation.

The necessary conditions are given in terms of weak Pontryagin principles, whereas they are given in terms of strong Pontryagin principles in Sect. 3.1.

Results of abstract optimization in ordered Banach spaces in the spirit of Karush–Kuhn–Tucker theorem are used to establish a Pontryagin maximum principle in the weak form. The used tools belong to linear and nonlinear functional analysis: sequence spaces, Nemytskii's operators, duality in topological spaces, and ordered Banach spaces. Notice that these tools cannot be used for problems with unbounded processes.

To compare our necessary conditions theorem with results established in [66], note that we do not need the convexity conditions on the criterion and on the constraints, and our assumptions are very different from Assumption 1 and Assumption 2 used in [66]; moreover we use vector states and vector controls and the authors of [66] use scalar states and controls.

In this section the controlled dynamical system is

(DI) $x_{t+1} \leq f(x_t, u_t)$

Let $\beta \in (0, 1)$ and $\eta \in \mathbb{R}^n_{++}$. In this section we set $X_t = \mathbb{R}^n_+$, $U_t = \mathbb{R}^d_+$. We denote by Adm^i_η the set of all processes $(\underline{x}, \underline{u}) \in (\mathbb{R}^n_+)^{\mathbb{N}} \times (\mathbb{R}^d_+)^{\mathbb{N}}$ which satisfy (DI) at each time $t \in \mathbb{N}$ and such that $x_0 = \eta$.

We formulate the following optimal control problem:

$$(\mathscr{B}_{\mathrm{i}}) \begin{cases} \text{Maximize } J(\underline{x}, \underline{u}) := \sum_{t=0}^{\infty} \beta^t \psi(x_t, u_t) \\ \quad \text{when} \quad x_0 = \eta \\ \qquad\qquad\quad \forall t \in \mathbb{N},\ x_{t+1} \leq f(x_t, u_t) \\ \qquad\qquad\quad \forall t \in \mathbb{N},\ x_t \geq 0,\ u_t \geq 0 \\ \qquad\qquad\quad (\underline{x}, \underline{u}) \in \ell^\infty(\mathbb{N}, \mathbb{R}^n) \times \ell^\infty(\mathbb{N}, \mathbb{R}^d) \end{cases}$$

which can be written

$(\mathscr{B}_{\mathrm{i}})$ Maximize $J(\underline{x}, \underline{u})$ when $(\underline{x}, \underline{u}) \in \mathrm{Adm}^i_\eta \cap \ell^\infty(\mathbb{N}, \mathbb{R}^n) \times \ell^\infty(\mathbb{N}, \mathbb{R}^d)$.

We consider the following list of conditions, where $\hat{x} \in \text{int}\ell^\infty(\mathbb{N}, \mathbb{R}^n)_+$ and $\hat{u} \in \text{int}\ell^\infty(\mathbb{N}, \mathbb{R}^d)_+$:

(H1) $\psi : \mathbb{R}^n \times \mathbb{R}^d \to \mathbb{R}$ and $f : \mathbb{R}^n \times \mathbb{R}^d \to \mathbb{R}^n$ are continuously Fréchet differentiable.
(H2) For all $t \in \mathbb{N}$, the partial differential with respect to the control variable $D_2 f(\hat{x}_t, \hat{u}_t)$ is positive and there exists $\gamma_1 \in (0, +\infty)$ such that, for all $t \in \mathbb{N}$, $D_2 f(\hat{x}_t, \hat{u}_t).1^{(d)} \geq \gamma_1^{(n)}$.
(H3) For all $t \in \mathbb{N}$, the partial differential with respect to the control variable $D_2 f(\hat{x}_t, \hat{u}_t)$ is negative and there exists $\gamma_2 \in (0, +\infty)$ such that, for all $t \in \mathbb{N}$, $D_2 f(\hat{x}_t, \hat{u}_t).(-1^{(d)}) \geq \gamma_2^{(n)}$.
(H4) For all $t \in \mathbb{N}$, the partial differential with respect to the control variable $D_2 f(\hat{x}_t, \hat{u}_t)$ is an isomorphism from $\mathbb{R}^d$ onto $\mathbb{R}^n$, and $\sigma := \sup_{t \in \mathbb{N}} \|D_2 f(\hat{x}_t, \hat{u}_t)^{-1}\| \in (0, +\infty)$.
(H5) There exists $\zeta \in (0, +\infty)$ such that, for all $t \in \mathbb{N}$, the partial differentials with respect to the state variable satisfy, for all $t \in \mathbb{N}$, $D_1 f(\hat{x}_t, \hat{u}_t).1^{(n)} \leq \zeta.1^{(n)}$.

The following theorem can be found in [26].

Theorem 3.3. *Under (H1), let $(\underline{\hat{x}}, \underline{\hat{u}}) \in \text{int}\,\ell^\infty(\mathbb{N}, \mathbb{R}^n)_+ \times \text{int}\,\ell^\infty(\mathbb{N}, \mathbb{R}^d)_+$ be a solution of problem $(\mathcal{B}_i)$. We also assume that (H2) or (H3) or (H4) or (H5) is fulfilled. Then there exists a sequence $(p_t)_{t \in \mathbb{N}_*} \in \ell^1(\mathbb{N}_*, \mathbb{R}^{n*})_+$ such that the following conditions are satisfied:*

(a) For all $t \in \mathbb{N}_*$, $p_t = p_{t+1} \circ D_1 f(\hat{x}_t, \hat{u}_t) + \beta^t D_1 \psi(\hat{x}_t, \hat{u}_t)$.
(b) For all $t \in \mathbb{N}$, $p_{t+1} \circ D_2 f(\hat{x}_t, \hat{u}_t) + \beta^t D_2 \psi(\hat{x}_t, \hat{u}_t) = 0$.
(c) For all $t \in \mathbb{N}$, $\langle p_{t+1}, f(\hat{x}_t, \hat{u}_t) - \hat{x}_{t+1} \rangle = 0$.
(d) $\lim_{t \to \infty} p_t = 0$.

The Pontryagin Hamiltonian is $H_t : \mathbb{R}^n_+ \times \mathbb{R}^d \times \mathbb{R} \times \mathbb{R}^{n*} \to \mathbb{R}$, defined by $H_t(x, u, p, \lambda_0) = \lambda_0 \beta^t \psi(x, u) + \langle p, f(x, u) \rangle$, the adjoint equation (a) is $p_t = D_1 H_t(\hat{x}_t, \hat{u}_t, p_{t+1}, 1)$, and the weak maximum principle (b) is $D_2 H_t(\hat{x}_t, \hat{u}_t, p_{t+1}, 1) = 0$. And the condition of complementary slackness (c) can be translated on the coordinates: $p^j_{t+1}.(f^j(\hat{x}_t, \hat{u}_t) - \hat{x}^j_{t+1}) = 0$ for all $j \in \{1, \ldots, n\}$, since $p_{t+1} \geq 0$ and $f(\hat{x}_t, \hat{u}_t) - \hat{x}_{t+1} \geq 0$. And so, $f^j(\hat{x}_t, \hat{u}_t) > \hat{x}^j_{t+1}$ implies $p^j_{t+1} = 0$. Moreover, since $(p_t)_t \in \ell^1(\mathbb{N}_*, \mathbb{R}^{n*})$, we necessarily have (d) $\lim_{t \to \infty} p_t = 0$ which is the transversality condition at infinity.

3.3.2 *Proof of Theorem 3.3*

The steps of the proof are the following:

First step: The optimal control problem can be written as the following abstract static optimization problem in an ordered Banach space:

$$\begin{cases} \text{Minimize } \mathscr{F}(\xi) \\ \text{when } \quad \xi \in S \end{cases}$$

We shall prove that Theorem B.11 in Appendix B can be used for our problem.

First we need to establish the following lemma:

Lemma 3.9. *Under (H1), the functional J, the Nemytskii's operator $N_f : \ell^\infty(\mathbb{N}, \mathbb{R}^n) \times \ell^\infty(\mathbb{N}, \mathbb{R}^d) \to \ell^\infty(\mathbb{N}, \mathbb{R}^n)$, defined by $N_f(\underline{x}, \underline{u}) := (f(x_t, u_t))_t$, and the operator $G : \ell^\infty(\mathbb{N}, \mathbb{R}^n) \times \ell^\infty(\mathbb{N}, \mathbb{R}^d) \to \ell^\infty(\mathbb{N}, \mathbb{R}^n)$, defined by $G(\underline{x}, \underline{u}) := (f(x_t, u_t) - x_{t+1})_t$, are of class C^1 on $\ell^\infty(\mathbb{N}, \mathbb{R}^n) \times \ell^\infty(\mathbb{N}, \mathbb{R}^d)$, and moreover the following formulas hold, for all $(\underline{x}, \underline{u})$, $(\underline{\delta x}, \underline{\delta u}) \in \ell^\infty(\mathbb{N}, \mathbb{R}^n) \times \ell^\infty(\mathbb{N}, \mathbb{R}^d)$:*

1. $DJ(\underline{x}, \underline{u}).(\underline{\delta x}, \underline{\delta u}) = \sum_{t=0}^{\infty} \beta^t D_1\psi(x_t, u_t)\delta x_t + \sum_{t=0}^{\infty} \beta^t D_2\psi(x_t, u_t)\delta u_t$.
2. $DN_f(\underline{x}, \underline{u})(\underline{\delta x}, \underline{\delta u}) = (D_1 f(x_t, u_t).\delta x_t)_t + (D_2 f(x_t, u_t).\delta u_t)_t$.
3. $DG(\underline{x}, \underline{u})(\underline{\delta x}, \underline{\delta u}) = (D_1 f(x_t, u_t)\delta x_t - \delta x_{t+1})_t + (D_2 f(x_t, u_t).\delta u_t)_t$.

This lemma is the same as Lemma 3.6. Let us verify now that all the hypotheses (1–7) of Theorem B.11Ê are satisfied. We set $\Xi := \ell^\infty(\mathbb{N}, \mathbb{R}^n) \times \ell^\infty(\mathbb{N}, \mathbb{R}^d)$; it is a Banach space as a product of two Banach spaces. We set $Y := \ell^\infty(\mathbb{N}, \mathbb{R}^n)$, $Z := \mathbb{R}^n$ and $\hat{\xi} := (\hat{\underline{x}}, \hat{\underline{u}})$.

We set $C := \ell^\infty(\mathbb{N}, \mathbb{R}^n)_+$; it is a convex cone and, using Proposition A.1 of Appendix A, its interior is nonempty, and so condition (1) of Theorem B.11 is fulfilled.

We set $\hat{S} := \{(\underline{x}, \underline{u}) \in \ell^\infty(\mathbb{N}, \mathbb{R}^n)_+ \times \ell^\infty(\mathbb{N}, \mathbb{R}^d)_+\}$; it is a convex set and, using Proposition A.1 of Appendix A, its interior is nonempty. And so condition (2) of Theorem B.11 is fulfilled.

We define $\mathscr{F} : \ell^\infty(\mathbb{N}, \mathbb{R}^n) \times \ell^\infty(\mathbb{N}, \mathbb{R}^d) \to \mathbb{R}$ by setting $\mathscr{F}(\underline{x}, \underline{u}) := -\sum_{t=0}^{\infty} \beta^t \psi(x_t, u_t)$; note that J is the restriction of $-\mathscr{F}$ to $\ell^\infty(\mathbb{N}, \mathbb{R}^n)_+ \times \ell^\infty(\mathbb{N}, \mathbb{R}^d)_+$. We define g as the restriction of $-G$ to $\ell^\infty(\mathbb{N}, \mathbb{R}^n)_+ \times \ell^\infty(\mathbb{N}, \mathbb{R}^d)_+$. Using Lemma 3.9 we can assert that J and g are of class C^1. And so the conditions (3) and (4) of Theorem B.11 are fulfilled.

We define $h : \ell^\infty(\mathbb{N}, \mathbb{R}^n) \times \ell^\infty(\mathbb{N}, \mathbb{R}^d) \to \mathbb{R}^n$ by setting $h(\underline{x}, \underline{u}) := -x_0 + \eta$. We define $k : \ell^\infty(\mathbb{N}, \mathbb{R}^n)_+ \times \ell^\infty(\mathbb{N}, \mathbb{R}^d)_+ \to \mathbb{R}^n$ as the restriction of $-h$ to $\ell^\infty(\mathbb{N}, \mathbb{R}^n)_+ \times \ell^\infty(\mathbb{N}, \mathbb{R}^d)_+$. We see that k is the restriction of an affine continuous functional, and consequently k is of class C^1. And so condition (5) of Theorem B.11 is fulfilled.

We set

$$S := \{(\underline{x}, \underline{u}) \in \hat{S} : (\forall t \in \mathbb{N},\ x_{t+1} \leq f(x_t, u_t)),\ x_0 = \eta\}.$$

This set is nonempty since it contains $(\hat{\underline{x}}, \hat{\underline{u}})$, and so condition (6) of Theorem B.11 is fulfilled. k is the restriction of an affine continuous functional and $Dk(\hat{\underline{x}}, \hat{\underline{u}})$ is clearly onto so the range of $Dh(\hat{\xi})$ is equal to $\mathbb{R}^n$ and so it is closed in $\mathbb{R}^n$. And so condition (7) of Theorem B.11 is fulfilled.

Second step: In this step we can apply Theorem B.11 and obtain the existence of $(\lambda_0, \Lambda_1, \Lambda_2)$ where $\lambda_0 \in [0, \infty)$, $\Lambda_1 \in \ell^\infty(\mathbb{N}, \mathbb{R}^n)^*$, Λ_1 is a positive functional, and $\Lambda_2 \in \mathscr{L}(\mathbb{R}^n, \mathbb{R})$ and such that

(i) $(\lambda_0, \Lambda_1, \Lambda_2)$ is nonzero.
(ii) $\langle \lambda_0 D\mathscr{F}(\hat{\xi}) + \Lambda_1 \circ Dg(\hat{\xi}) + \Lambda_2 \circ Dh(\hat{\xi}), \xi - \hat{\xi} \rangle \geq 0$ for all $\xi = (\underline{x}, \underline{u}) \in \ell^\infty(\mathbb{N}, \mathbb{R}^n) \times \ell^\infty(\mathbb{N}, \mathbb{R}^d)$ such that $x_t \in \mathbb{R}^n_{++}$ and $u_t \in \mathbb{R}^d_{++}$ for all $t \in \mathbb{N}$.
(iii) $\langle \Lambda_1, g(\hat{\xi}) \rangle = 0$.

Using Proposition A.3 and Theorem A.3 in Appendix A, we know that there exist $\underline{q} \in \ell^1(\mathbb{N}, \mathbb{R}^{n*})_+$ and $\theta \in \ell^1_d(\mathbb{N}, \mathbb{R}^n)$ positive such that $\Lambda_1 = \underline{q} + \theta$.

To use conclusion (ii) we need the following lemma.

Lemma 3.10. *Let $\zeta \in \mathbb{R}^k$ and $z \in \mathbb{R}^k_{++}$ such that $\langle \zeta, w - z \rangle \leq 0$ for all $w \in \mathbb{R}^k_{++}$. Then we have $\zeta = 0$.*

Proof. We set $\varrho := \min\{z^i : i \in \{1, \dots, k\}\} > 0$ and $B(0, \varrho) := \{v \in \mathbb{R}^k : |v|_\infty < \varrho\}$. Let $v \in B(0, \varrho)$; setting $w := v + z$ we have $w \in \mathbb{R}^k_{++}$. Since $v = w - z$ we have proven that $\langle \zeta, v \rangle \leq 0$ for all $v \in B(0, \varrho)$, and using the symmetry of $B(0, \varrho)$ we obtain $\langle \zeta, v \rangle = 0$ for all $v \in B(0, \varrho)$. When $y \in \mathbb{R}^k \setminus B(0, \varrho)$, we have $\frac{\varrho}{2|y|_\infty} y \in B(0, \varrho)$, and then we obtain $0 = \langle \zeta, (\frac{\varrho}{2|y|_\infty} y) \rangle = \frac{\varrho}{2|y|_\infty} \langle \zeta, y \rangle$ which implies $\langle \zeta, y \rangle = 0$. □

Using (ii) and the formulas of operators given in Lemma 3.9, we obtain the following relation, for all $(\underline{x}, \underline{u}) \in \ell^\infty(\mathbb{N}, \mathbb{R}^n) \times \ell^\infty(\mathbb{N}, \mathbb{R}^d)$ such that $x_t \in \mathbb{R}^n_{++}$ and $u_t \in \mathbb{R}^d_{++}$ for all $t \in \mathbb{N}$:

$$\left.\begin{aligned} 0 \geq \lambda_0 . \sum_{t=0}^{\infty} \beta^t D\psi(\hat{x}_t, \hat{u}_t)(x_t - \hat{x}_t, u_t - \hat{u}_t) \\ + \langle \Lambda_1, (Df(\hat{x}_t, \hat{u}_t)(x_t - \hat{x}_t, u_t - \hat{u}_t) - (x_{t+1} - \hat{x}_{t+1}))_t \rangle \\ + \Lambda_2(x_0 - \hat{x}_0). \end{aligned}\right\} \tag{3.10}$$

To abridge the writing, we introduce, for all $t \in \mathbb{N}$,

$$d_t := Df(\hat{x}_t, \hat{u}_t)(x_t - \hat{x}_t, u_t - \hat{u}_t) - (x_{t+1} - \hat{x}_{t+1}). \tag{3.11}$$

We arbitrarily fix $t \in \mathbb{N}_*$, we fix $\underline{u} = \underline{\hat{u}}$, and we fix $x_s = \hat{x}_s$ when $s \in \mathbb{N} \setminus \{t\}$, and x_t varies inside $\mathbb{R}^n_{++}$. Then we have $d_s = 0$ when $s \notin \{t-1, t\}$, $d_{t-1} = -(x_t - \hat{x}_t)$ and $d_t = D_1 f(\hat{x}_t, \hat{u}_t)(x_t - \hat{x}_t)$. And so we have $\underline{d} \in c_0(\mathbb{N}, \mathbb{R}^n)$ and consequently $\theta(\underline{d}) = \langle c, \lim_{s \to \infty} d_s \rangle = \langle c, 0 \rangle = 0$ (see Definition A.2 in Appendix) and $\langle \underline{q}, \underline{d} \rangle = \langle q_{t-1}, -(x_t - \hat{x}_t) \rangle + \langle q_t, D_1 f(\hat{x}_t, \hat{u}_t)(x_t - \hat{x}_t) \rangle = \langle -q_{t-1} + q_t \circ D_1 f(\hat{x}_t, \hat{u}_t), x_t - \hat{x}_t \rangle$. And so we have proven the following relation, for all $x_t \in \mathbb{R}^n_{++}$:

$$\langle \Lambda_1, \underline{d} \rangle = \langle -q_{t-1} + q_t \circ D_1 f(\hat{x}_t, \hat{u}_t), x_t - \hat{x}_t \rangle.$$

Moreover we have $\lambda_0.\sum_{t=0}^{\infty}\beta^t D\psi(\hat{x}_t,\hat{u}_t).(x_t-\hat{x}_t,u_t-\hat{u}_t)=\lambda_0.\beta^t.D_1\psi(\hat{x}_t,\hat{u}_t).(x_t-\hat{x}_t)$. Note that $\Lambda_2(x_0-\hat{x}_0)=\Lambda_2(0)=0$. And then, from (3.10) and from the previous calculations, we obtain the following relation, for all $t\in\mathbb{N}_*$, for all $x_t\in\mathbb{R}^n_{++}$:

$$\langle\lambda_0.\beta^t D_1\psi(\hat{x}_t,\hat{u}_t)-q_{t-1}+q_t\circ D_1 f(\hat{x}_t,\hat{u}_t),x_t-\hat{x}_t\rangle\leq 0,$$

and using Lemma 3.10 we obtain

$$-q_{t-1}+q_t\circ D_1 f(\hat{x}_t,\hat{u}_t)+\lambda_0.\beta^t D_1\psi(\hat{x}_t,\hat{u}_t)=0. \tag{3.12}$$

Now we take $\underline{x}=\hat{\underline{x}}$, we arbitrarily fix $t\in\mathbb{N}$, and we take $u_s=\hat{u}_s$ when $s\in\mathbb{N}\setminus\{t\}$, and u_t varies inside $\mathbb{R}^d_{++}$. Then $\lambda_0.\sum_{t=0}^{\infty}\beta^t D\psi(\hat{x}_t,\hat{u}_t)(x_t-\hat{x}_t,u_t-\hat{u}_t)=\lambda_0.\beta^t.D_2\psi(\hat{x}_t,\hat{u}_t)(u_t-\hat{u}_t)$. When $s\neq t$ we have $d_s=Df(\hat{x}_s,\hat{u}_s)(0,0)-0=0$ and $d_t=D_2 f(\hat{x}_t,\hat{u}_t)(u_t-\hat{u}_t)$. And so we have $\underline{d}\in c_0(\mathbb{N},\mathbb{R}^n)$ which implies $\theta(\underline{d})=0$, and $\langle\underline{q},\underline{d}\rangle=\langle q_t,d_t\rangle=\langle q_t,D_2 f(\hat{x}_t,\hat{u}_t)(u_t-\hat{u}_t)\rangle=\langle q_t\circ D_2 f(\hat{x}_t,\hat{u}_t),u_t-\hat{u}_t\rangle$. Then from (3.10) we deduce the following relation, for all $u_t\in\mathbb{R}^d_{++}$:

$$0\geq\langle\lambda_0.\beta^t D_2\psi(\hat{x}_t,\hat{u}_t)+q_t\circ D_2 f(\hat{x}_t,\hat{u}_t),u_t-\hat{u}_t\rangle,$$

and using Lemma 3.10 we obtain the following relation, for all $t\in\mathbb{N}$:

$$\lambda_0.\beta^t D_2\psi(\hat{x}_t,\hat{u}_t)+q_t\circ D_2 f(\hat{x}_t,\hat{u}_t)=0. \tag{3.13}$$

Now we use the conclusion (iii). We set $w_t:=f(\hat{x}_t,\hat{u}_t)-\hat{x}_{t+1}\in\mathbb{R}^n_+$. After (iii), we have $0=\Lambda_1(\underline{w})=\langle\underline{q},\underline{w}\rangle+\theta(\underline{w})$. Using Theorem A.3 of Appendix, since $\underline{w}\geq\underline{0}$, we have $\langle\underline{q},\underline{w}\rangle\geq 0$ and $\theta(\underline{w})\geq 0$, and since their sum is equal to zero, we obtain $\langle\underline{q},\underline{w}\rangle=\overline{0}$ and $\theta(\underline{w})=0$. Since $q_t\geq 0$ and $w_t\geq 0$, we have $\langle q_t,w_t\rangle\geq 0$ for all $t\in\mathbb{N}$. And then the relation $0=\langle\underline{q},\underline{w}\rangle=\sum_{t=0}^{\infty}\langle q_t,w_t\rangle$ implies that $\langle q_t,w_t\rangle=0$ for all $t\in\mathbb{N}$. And so, we have proven, for all $t\in\mathbb{N}$, the following relation:

$$\langle q_t,f(\hat{x}_t,\hat{u}_t)-\hat{x}_{t+1}\rangle=0. \tag{3.14}$$

Third step: In this step we prove that (H2) implies hypothesis (Q2) of Theorem B.11.

First translating (Q2) in terms of problem $(\mathscr{P})$ we obtain the following condition:

$$\left.\begin{array}{r}\exists(\tilde{\underline{x}},\tilde{\underline{u}})\in\mathrm{int}(\ell^{\infty}(\mathbb{N},\mathbb{R}^n)_+\times\ell^{\infty}(\mathbb{N},\mathbb{R}^d)_+)\ \text{s.t.}\\ (f(\hat{x}_t,\hat{u}_t)-\hat{x}_{t+1}+Df(\hat{x}_t,\hat{u}_t)(\tilde{x}_t-\hat{x}_t,\tilde{u}_t-\hat{u}_t)-(\tilde{x}_{t+1}-\hat{x}_{t+1}))_{t\in\mathbb{N}}\\ \in\mathrm{int}\,\ell^{\infty}(\mathbb{N},\mathbb{R}^n)_+,\text{and}\ \ Dh(\hat{x},\hat{u})(\tilde{x}-\hat{x},\tilde{u}-\hat{u})=0.\end{array}\right\} \tag{3.15}$$

The last condition of (7) is simply

$$\tilde{x}_0=\hat{x}_0=\eta. \tag{3.16}$$

Since $\underline{\hat{u}}$ belongs to the interior of $\ell^{\infty}(\mathbb{N}, \mathbb{R}^d)_+$, using Lemma A.1, we know that there exists $\alpha \in (0, \infty)$ such that, for all $t \in \mathbb{N}$, $\hat{u}_t \geq \alpha^{(d)}$. We set $\underline{\tilde{x}} = \underline{\hat{x}}$; and so $\underline{\tilde{x}}$ belongs to the interior of $\ell^{\infty}(\mathbb{N}, \mathbb{R}^d)_+$. We set $\tilde{u}_t = r.\hat{u}_t$ where $r \in (1, \infty)$ for all $t \in \mathbb{N}$. Then we have $\tilde{u}_t \geq r.\alpha^{(d)} = (r.\alpha)^{(d)}$ that implies which $\underline{\tilde{u}}$ belongs to the interior of $\ell^{\infty}(\mathbb{N}, \mathbb{R}^d)_+$ after Lemma A.1. Moreover, for all $t \in \mathbb{N}$, we have $D_2 f(\hat{x}_t, \hat{u}_t)(\tilde{u}_t - \hat{u}_t) = (r-1).D_2 f(\hat{x}_t, \hat{u}_t)(\hat{u}_t) \geq (r-1).D_2 f(\hat{x}_t, \hat{u}_t)(\alpha^{(d)}) = (r-1).\alpha.D_2 f(\hat{x}_t, \hat{u}_t)(1^{(d)}) \geq (r-1).\alpha.\gamma_1^{(n)}$ using (H2). Using $f(\hat{x}_t, \hat{u}_t) - \hat{x}_{t+1} \geq 0$, we have $f(\hat{x}_t, \hat{u}_t) - \hat{x}_{t+1} + Df(\hat{x}_t, \hat{u}_t)(\tilde{x}_t - \hat{x}_t, \tilde{u}_t - \hat{u}_t) - (\tilde{x}_{t+1} - \hat{x}_{t+1}) \geq D_2 f(\hat{x}_t, \hat{u}_t)(\tilde{u}_t - \hat{u}_t) \geq (r-1).\alpha.\gamma_1^{(n)}$ which belongs to the interior of $\ell^{\infty}(\mathbb{N}, \mathbb{R}^n)_+$. And this implies that the condition (3.15) is fulfilled and consequently (Q2) is fulfilled.

Fourth step: In this step, we prove that (H3) implies hypothesis (Q2) of Theorem B.11.

The proof is similar to that of the third step but with $r \in (0, 1)$ instead of $(1, +\infty)$ which implies that we have $D_2 f(\hat{x}_t, \hat{u}_t)(\tilde{u}_t - \hat{u}_t) \geq (1-r).D_2 f(\hat{x}_t, \hat{u}_t)(-\alpha^{(d)})$ (since $D_2 f(\hat{x}_t, \hat{u}_t)$ is negative) $= (1-r).\alpha.D_2 f(\hat{x}_t, \hat{u}_t)(-1^{(d)}) \geq (1-r).\alpha.\gamma_2^{(n)}$. The conclusion follows as in the third step.

Fifth step: In this step, we prove that (H4) implies hypothesis (Q2) of Theorem B.11.

We set $\underline{\tilde{x}} = \underline{\hat{x}}$. As previously we know that $\hat{u}_t \geq \alpha^{(d)}$ for all $t \in \mathbb{N}$ where $\alpha \in (0, +\infty)$. We introduce $k_t := \frac{1}{2}(\frac{\alpha}{\sigma})^{(n)}$ for all $t \in \mathbb{N}$. Then we have $\underline{k} \in \operatorname{int} \ell^{\infty}(\mathbb{N}, \mathbb{R}^n)_+$. We define $\tilde{u}_t := \hat{u}_t + D_2 f(\hat{x}_t, \hat{u}_t)^{-1}(k_t)$ for all $t \in \mathbb{N}$. Note that, under (H4), we have $n = d$. Then we have

$$\begin{aligned}
\tilde{u}_t &\geq \alpha^{(n)} + D_2 f(\hat{x}_t, \hat{u}_t)^{-1}(k_t) \\
&= \alpha^{(n)} + D_2 f(\hat{x}_t, \hat{u}_t)^{-1}(\tfrac{1}{2}(\tfrac{\alpha}{\sigma})^{(n)}) \\
&= \alpha^{(n)} + \frac{1}{2}.\frac{\alpha}{\sigma}.D_2 f(\hat{x}_t, \hat{u}_t)^{-1}(1^{(n)}) \\
&\geq \alpha^{(n)} + \frac{1}{2}.\frac{\alpha}{\sigma}.(-|D_2 f(\hat{x}_t, \hat{u}_t)^{-1}(1^{(n)})|_{\infty}^{(n)}) \\
&\geq \alpha^{(n)} + \frac{1}{2}.\frac{\alpha}{\sigma}.(-\|D_2 f(\hat{x}_t, \hat{u}_t)^{-1}\|.|1^{(n)}|_{\infty})^{(n)} \\
&\geq \alpha^{(n)} + \frac{1}{2}.\frac{\alpha}{\sigma}.(-\sigma)^{(n)} = \alpha^{(n)} - \tfrac{1}{2}.\alpha^{(n)} = \frac{1}{2}.\alpha^{(n)},
\end{aligned}$$

which implies that $\underline{\tilde{u}} \in \operatorname{int} \ell^{\infty}(\mathbb{N}, \mathbb{R}^n)_+$.

Using $f(\hat{x}_t, \hat{u}_t) - \hat{x}_{t+1} \geq 0$, we have $(f(\hat{x}_t, \hat{u}_t) - \hat{x}_{t+1} + Df(\hat{x}_t, \hat{u}_t)(\tilde{x}_t - \hat{x}_t, \tilde{u}_t - \hat{u}_t) - (\tilde{x}_{t+1} - \hat{x}_{t+1}))_t \geq (D_2 f(\hat{x}_t, \hat{u}_t)(\tilde{u}_t - \hat{u}_t))_t = (k_t)_t$ which belongs to $\operatorname{int}\ell^{\infty}(\mathbb{N}, \mathbb{R}^n)_+$.

Sixth step: In this step, we prove that (H5) implies hypothesis (Q2) of Theorem B.11.

We set $\underline{\tilde{u}} = \underline{\hat{u}}$. Using Lemma A.1, since $\underline{\hat{x}}$ is interior to $\ell^\infty(\mathbb{N}, \mathbb{R}^n)_+$, there exists $a \in (0, +\infty)$ such that $\hat{x}_t \geq a^{(n)}$. We fix $b \in (0, 1)$ and we set $\tilde{x}_0 := \hat{x}_0$ and $\tilde{x}_t := \hat{x}_t - b.a^{(n)}$ when $t \in \mathbb{N}_*$. Then we have $\tilde{x}_t \geq a^{(n)} - b.a^{(n)} = (1-b)a^{(n)}$ for all $t \in \mathbb{N}_*$ which implies that $\underline{\tilde{x}}$ belongs to the interior of $\ell^\infty(\mathbb{N}, \mathbb{R}^n)_+$.

When $t = 0$, $D_1 f(\hat{x}_0, \hat{u}_0)(\tilde{x}_0 - \hat{x}_0) - (\tilde{x}_1 - \hat{x}_1) = -(\tilde{x}_1 - \hat{x}_1) = b.a^{(n)}$.

When $t \in \mathbb{N}_*$, $D_1 f(\hat{x}_t, \hat{u}_t)(\tilde{x}_t - \hat{x}_t) - (\tilde{x}_{t+1} - \hat{x}_{t+1}) = D_1 f(\hat{x}_t, \hat{u}_t)(-b.a^{(n)}) + b.a^{(n)} = -b.a.D_1 f(\hat{x}_t, \hat{u}_t)(1^{(n)}) + b.a.1^{(n)} \geq -b.a.\zeta.1^{(n)} + b.a.1^{(n)} = b.a.(1 - \zeta).1^{(n)}$. Using $f(\hat{x}_t, \hat{u}_t) - \hat{x}_{t+1} \geq 0$, we have $(f(\hat{x}_t, \hat{u}_t) - \hat{x}_{t+1} + Df(\hat{x}_t, \hat{u}_t)(\tilde{x}_t - \hat{x}_t, \tilde{u}_t - \hat{u}_t) - (\tilde{x}_{t+1} - \hat{x}_{t+1}))_t \geq (D_1 f(\hat{x}_t, \hat{u}_t)(\tilde{x}_t - \hat{x}_t) - (\tilde{x}_{t+1} - \hat{x}_{t+1}))_t$ which belongs to $\mathrm{int}\ell^\infty(\mathbb{N}, \mathbb{R}^n)_+$. And so condition (3.15) is fulfilled which implies that (Q2) is satisfied.

Last step: Note that hypothesis (Q1) of Theorem B.11. is clearly satisfied. Using the third, fourth, fifth, and sixth steps, we can take $\lambda_0 \neq 0$ into the conclusions (i), (ii), and (iii) after the first two steps, and since the set of all $(\lambda_0, \Lambda_1, \Lambda_2)$ which satisfies the conclusions (i), (ii), and (iii) is a cone, we can choose $\lambda_0 = 1$. Then setting $p_t := q_{t-1}$ when $t \in \mathbb{N}_*$, (3.12) becomes (a), (3.13) becomes (b), and (3.14) becomes (c). And so Theorem 3.3 is proven.

3.4 Links with Unbounded Problems

We define $\mathrm{Dom}^e_\eta(J) = \{(\underline{x}, \underline{u}) \in \mathrm{Adm}^e_\eta : \sum_{t=0}^{+\infty} \beta^t \psi(x_t, u_t) \text{ converges in } \mathbb{R}\}$.

And we define $\mathrm{Dom}^i_\eta(J) = \{(\underline{x}, \underline{u}) \in \mathrm{Adm}^i_\eta : \sum_{t=0}^{+\infty} \beta^t \psi(x_t, u_t) \text{ converges in } \mathbb{R}\}$.

And so we can consider the following problems that we have considered in Chap. 1, when $a \in \{e, i\}$:

$(\mathcal{P}^n_a)$ Maximize $J(\underline{x}, \underline{u})$ when $(\underline{x}, \underline{u}) \in \mathrm{Dom}^a_\eta(J)$.

$(\mathcal{P}^s_a)$ Find $(\underline{\hat{x}}, \underline{\hat{u}}) \in \mathrm{Dom}^a_\eta(J)$ such that, for all $(\underline{x}, \underline{u}) \in \mathrm{Adm}^a_\eta$

$$J(\underline{\hat{x}}, \underline{\hat{u}}) \geq \limsup_{T \to \infty} \sum_{t=0}^{T} \beta^t \psi(x_t, u_t).$$

$(\mathcal{P}^o_a)$ Find $(\underline{\hat{x}}, \underline{\hat{u}}) \in \mathrm{Adm}^a_\eta$ such that, for all $(\underline{x}, \underline{u}) \in \mathrm{Adm}^a_\eta$

$$\liminf_{T \to \infty} \left(\sum_{t=0}^{T} \beta^t \psi(\hat{x}_t, \hat{u}_t) - \sum_{t=0}^{T} \beta^t \psi(x_t, u_t) \right) \geq 0.$$

$(\mathcal{P}^w_a)$ Find $(\underline{\hat{x}}, \underline{\hat{u}}) \in \mathrm{Adm}^a_\eta$ such that, for all $(\underline{x}, \underline{u}) \in \mathrm{Adm}^a_\eta$

$$\limsup_{T \to \infty} \left(\sum_{t=0}^{T} \beta^t \psi(\hat{x}_t, \hat{u}_t) - \sum_{t=0}^{T} \beta^t \psi(x_t, u_t) \right) \geq 0.$$

The following theorems give conditions under which the existence of solutions of bounded problems implies the existence of solutions of unbounded problems. They show that under a nonnegativity assumption, solving the problem in the space of bounded processes provides solutions for problems in spaces of admissible processes which are not necessarily bounded. They can be found in [24].

Theorem 3.4. *We assume the following conditions fulfilled:*

(i) $\psi \geq 0$ *on* $\Omega \times U$.
(ii) *For all* $t \in \mathbb{N}$, *for all* $x_t \in \Omega$, *there exists* $u_t \in U_t$ such that $x_t = f(x_t, u_t)$.

Then we have

(a) $\sup_{(\underline{x},\underline{u})\in \mathrm{Dom}^e_\eta(J)} J(\underline{x},\underline{u}) = \sup_{(\underline{x},\underline{u})\in \mathrm{Adm}^e_\eta \cap (\mathscr{X}\times\mathscr{U})} J(\underline{x},\underline{u})$.
(b) *If* $(\hat{\underline{x}}, \hat{\underline{u}})$ *is a solution of problem* $(\mathscr{B}_e)$, *then it is a solution of problems* $(\mathscr{P}^s_e)$, $(\mathscr{P}^o_e)$ *and* $(\mathscr{P}^w_e)$ *which all reduce to the same problem.*

Theorem 3.5. *We assume the following conditions fulfilled:*

(i') $\psi \geq 0$ *on* $\mathbb{R}^n \times \mathbb{R}^d$.
(ii') *For all* $x \in \mathbb{R}^n_+$, *there exists* $u \in \mathbb{R}^d_+$ such that $x \leq f(x, u)$.

Then we have

(a') $\sup_{(\underline{x},\underline{u})\in \mathrm{Dom}^i_\eta(J)} J(\underline{x},\underline{u}) = \sup_{(\underline{x},\underline{u})\in \mathrm{Adm}^i_\eta \cap (\ell^\infty(\mathbb{N},\mathbb{R}^n)\times\ell^\infty(\mathbb{N},\mathbb{R}^d))} J(\underline{x},\underline{u})$.
(b') *If* $(\hat{\underline{x}}, \hat{\underline{u}})$ *is a solution of problem* $(\mathscr{B}_i)$, *then it is a solution of problems* $(\mathscr{P}^s_i)$, $(\mathscr{P}^o_i)$ *and* $(\mathscr{P}^w_i)$ *which all reduce to the same problem.*

Proof. We give the proof of Theorem 3.4 which is in (see [24]). The proof of Theorem 3.5 is analogous.

(a) It is clear that the following inequality holds:

$$\sup_{(\underline{x},\underline{u})\in \mathrm{Dom}^e_\eta(J)} J(\underline{x},\underline{u}) \geq \sup_{(\underline{x},\underline{u})\in (\mathscr{X}\times\mathscr{U})\cap \mathrm{Adm}^e_\eta} J(\underline{x},\underline{u})$$

Let $(\tilde{\underline{x}}, \tilde{\underline{u}}) \in \mathrm{Dom}^e_\eta(J)$. Let $\epsilon > 0$ be given and let $T = T_\epsilon$ be such that $0 \leq \sum_{t>T_\epsilon} \beta^t \psi(\tilde{x}_t, \tilde{u}_t) \leq \epsilon$. Set

$$x'_t = \begin{cases} \tilde{x}_t & \text{if } t \leq T \\ \tilde{x}_T & \text{if } t > T \end{cases}$$

and

$$u'_t = \begin{cases} \tilde{u}_t & \text{if } t \leq T \\ u'_T & \text{if } t > T \end{cases}$$

where u'_T is such that $x'_T = f(x'_T, u'_T)$.
$\underline{x}' = (x'_t)_t$ and $\underline{u}' = (u'_t)_t$ are bounded and $(\underline{x}', \underline{u}') \in (\mathscr{X} \times \mathscr{U}) \cap \mathrm{Adm}^e_\eta$.
Since $\psi \geq 0$ on $\Omega \times U$ we have

$$J(\underline{x}',\underline{u}') = \sum_{t=0}^{+\infty} \beta^t \psi(x'_t, u'_t) = \sum_{t=0}^{T} \beta^t \psi(\tilde{x}_t, \tilde{u}_t) + \sum_{t=T+1}^{+\infty} \beta^t \psi(\tilde{x}_T, u'_T) \geq \sum_{t=0}^{T} \beta^t \psi(\tilde{x}_t, \tilde{u}_t)$$

$$\sup_{(\underline{x},\underline{u})\in(\mathscr{X}\times\mathscr{U})\cap \mathrm{Adm}^e_\eta} J(\underline{x},\underline{u}) \geq J(\underline{x}',\underline{u}') \geq \sum_{t=0}^{T} \beta^t \psi(\tilde{x}_t, \tilde{u}_t)$$

so we obtain

$$\epsilon + \sup_{(\underline{x},\underline{u})\in(\mathscr{X}\times\mathscr{U})\cap \mathrm{Adm}^e_\eta} J(\underline{x},\underline{u}) \geq J(\underline{\tilde{x}},\underline{\tilde{u}}).$$

Since this is true for all $\epsilon > 0$, letting $\epsilon \to 0$, we obtain

$$\sup_{(\underline{x},\underline{u})\in(\mathscr{X}\times\mathscr{U})\cap \mathrm{Adm}^e_\eta} J(\underline{x},\underline{u}) \geq \sup_{(\underline{x},\underline{u})\in \mathrm{Dom}^e_\eta(J)} J(\underline{x},\underline{u})$$

(b) Since $\psi \geq 0$, for all $(\underline{x},\underline{u}) \in \mathrm{Adm}^e_\eta$, the sequence $(\sum_{t=0}^{T} \beta^t \psi(x_t, u_t))_T$ is nonnegative and nondecreasing so it converges in $[0, \infty]$.

So $\limsup_{T\to\infty} \sum_{t=0}^{T} \beta^t \psi(x_t, u_t) = \lim_{T\to\infty} \sum_{t=0}^{T} \beta^t \psi(x_t, u_t)$. Hence $(\mathscr{P}^s_e)$ and $(\mathscr{P}^o_e)$ reduce to the same problem. Similarly $(\mathscr{P}^w_e)$ reduces to it. Let $(\underline{\hat{x}},\underline{\hat{u}})$ be a solution of problem $(\mathscr{P}^b_e)$ and suppose it is not a solution of problem $(\mathscr{P}^s_e)$. So there exists $(\underline{x},\underline{u}) \in \mathrm{Adm}^e_\eta$ such that $\lim_{T\to\infty} \sum_{t=0}^{T} \beta^t \psi(x_t, u_t) = +\infty$ that is

$$\forall R \in \mathbb{R}, \exists T_R \in \mathbb{N}_*, \forall T \geq T_R, \sum_{t=0}^{T} \beta^t \psi(x_t, u_t) > R.$$

Let $R \in \mathbb{R}$ and $T = T_R$. Construct $\underline{x}' = (x'_t)_t$ and $\underline{u}' = (u'_t)_t$ as in (a). Thus

$$\sum_{t=0}^{+\infty} \beta^t \psi(x'_t, u'_t) = \sum_{t=0}^{T} \beta^t \psi(x_t, u_t) + \sum_{t=T+1}^{+\infty} \beta^t \psi(x_T, u_T) \geq R.$$

Hence we obtain $\sup_{(\underline{x},\underline{u})\in(\mathscr{X}\times\mathscr{U})\cap \mathrm{Adm}^e_\eta} J(\underline{x},\underline{u}) \geq R$, so $\sup_{(\underline{x},\underline{u})\in(\mathscr{X}\times\mathscr{U})\cap \mathrm{Adm}^e_\eta} J(\underline{x},\underline{u}) = +\infty$ which contradicts the hypothesis, so $(\underline{\hat{x}},\underline{\hat{u}})$ is a solution of problem $(\mathscr{P}^s_e)$. □

3.5 Sufficient Conditions

Theorem 3.6. *Let U_t be convex for every t. Let $(\hat{\underline{x}}, \hat{\underline{u}}) \in \mathscr{X} \times \mathscr{U} \cap \mathrm{Adm}^e_\eta$. We assume that there exists $(p_t)_{t \in \mathbb{N}_*} \in \ell^1(\mathbb{N}_*, \mathbb{R}^{n*})$ and that the following conditions are fulfilled:*

(i) *The mappings ψ and f are of class C^1 on $\Omega \times U$.*
(ii) For all $t \in \mathbb{N}_*$, $p_t = p_{t+1} \circ D_1 f(\hat{x}_t, \hat{u}_t) + \beta^t D_1 \psi(\hat{x}_t, \hat{u}_t)$.
(iii) For all $t \in \mathbb{N}$, for all $u_t \in U_t$,
$$\beta^t \psi(\hat{x}_t, \hat{u}_t) + \langle p_{t+1}, f(\hat{x}_t, \hat{u}_t)\rangle \geq \beta^t \psi(\hat{x}_t, u_t) + \langle p_{t+1}, f(\hat{x}_t, u_t)\rangle.$$
(iv) For all $t \in \mathbb{N}$, *the mapping H_t is concave with respect to (x_t, u_t).*

Then $(\hat{\underline{x}}, \hat{\underline{u}})$ is a solution of $(\mathscr{B}_{\mathbf{e}})$.

Proof. Let $(\underline{x}, \underline{u}) \in \mathscr{X} \times \mathscr{U} \cap \mathrm{Adm}^e_\eta$. Recall that the adjoint equation (ii) is $p_t = D_1 H_t(\hat{x}_t, \hat{u}_t, p_{t+1}, 1)$ and the strong maximum principle (iii) is $H_t(\hat{x}_t, \hat{u}_t, p_{t+1}, 1) \geq H_t(\hat{x}_t, u_t, p_{t+1}, 1)$. For all $t \in \mathbb{N}$, we have

$$\begin{aligned}
&\beta^t \psi(\hat{x}_t, \hat{u}_t) - \beta^t \psi(x_t, u_t)\\
&= H_t(\hat{x}_t, \hat{u}_t, p_{t+1}, 1) - H_t(x_t, u_t, p_{t+1}, 1) - \langle p_{t+1}, f(\hat{x}_t, \hat{u}_t) - f(x_t, u_t)\rangle\\
&= H_t(\hat{x}_t, \hat{u}_t, p_{t+1}, 1) - H_t(x_t, u_t, p_{t+1}, 1) - \langle D_1 H_{t+1}(\hat{x}_{t+1}, \hat{u}_{t+1}, p_{t+2}, 1), \hat{x}_{t+1}\\
&\quad - x_{t+1}\rangle - \langle D_2 H_t(\hat{x}_t, \hat{u}_t, p_{t+1}, 1), \hat{u}_t - u_t\rangle + \langle D_2 H_t(\hat{x}_t, \hat{u}_t, p_{t+1}, 1), \hat{u}_t - u_t\rangle,
\end{aligned}$$

Therefore, using $\hat{x}_0 = x_0$, we have

$$\begin{aligned}
&\sum_{t=0}^{T} (\beta^t \psi(\hat{x}_t, \hat{u}_t) - \beta^t \psi(x_t, u_t))\\
&= \sum_{t=0}^{T} (H_t(\hat{x}_t, \hat{u}_t, p_{t+1}, 1) - H_t(x_t, u_t, p_{t+1}, 1) - \langle D_1 H_t(\hat{x}_t, \hat{u}_t, p_{t+1}, 1), \hat{x}_t - x_t\rangle\\
&\quad - \langle D_2 H_t(\hat{x}_t, \hat{u}_t, p_{t+1}, 1), \hat{u}_t - u_t\rangle) - \langle p_{T+1}, \hat{x}_{T+1} - x_{T+1}\rangle\\
&\quad + \sum_{t=0}^{T} \langle D_2 H_t(\hat{x}_t, \hat{u}_t, p_{t+1}, 1), \hat{u}_t - u_t\rangle.
\end{aligned}$$

Since for all $t \in \mathbb{N}$ H_t is concave with respect to (x_t, u_t), we have for all $t \in \mathbb{N}$,

$$\begin{aligned}
&H_t(\hat{x}_t, \hat{u}_t, p_{t+1}, 1) - H_t(x_t, u_t, p_{t+1}, 1) - \langle D_1 H_t(\hat{x}_t, \hat{u}_t, p_{t+1}, 1), \hat{x}_t - x_t\rangle\\
&\quad - \langle D_2 H_t(\hat{x}_t, \hat{u}_t, p_{t+1}, 1), \hat{u}_t - u_t\rangle \geq 0.
\end{aligned}$$

Hypothesis (iii) implies $\langle D_2 H_t(\hat{x}_t, \hat{u}_t, p_{t+1}, 1), \hat{u}_t - u_t\rangle \geq 0$ which is the first order necessary condition for the optimality of $\hat{u}_t$. Thus we have

$$\sum_{t=0}^{T}(\beta^t\psi(\hat{x}_t,\hat{u}_t)-\beta^t\psi(x_t,u_t))\geq\langle p_{T+1},x_{T+1}-\hat{x}_{T+1}\rangle.$$

The hypothesis $(p_t)_{t\in\mathbb{N}_*}\in\ell^1(\mathbb{N}_*,\mathbb{R}^{n*})$ implies $\lim_{t\longrightarrow+\infty}p_t=0$ and since $\underline{\hat{x}}$ and $\underline{x}$ belong to $\ell^\infty(\mathbb{N},\mathbb{R}^n)$ we have $||\underline{x}-\underline{\hat{x}}||\leq||\underline{x}||+||\underline{\hat{x}}||<\infty$. Hence we obtain $\lim_{T\longrightarrow+\infty}\langle p_{T+1},x_{T+1}-\hat{x}_{T+1}\rangle=0$ so $\lim_{T\longrightarrow+\infty}\sum_{t=0}^{T}(\beta^t\psi(\hat{x}_t,\hat{u}_t)-\beta^t\psi(x_t,u_t))\geq 0$. That is $J(\underline{\hat{x}},\underline{\hat{u}})-J(\underline{x},\underline{u})\geq 0$. Hence $(\underline{\hat{x}},\underline{\hat{u}})$ is a solution of $(\mathscr{B}_{\mathbf{e}})$.

Remark. (iii) could be replaced by

$$\langle D_2H_t(\hat{x}_t,\hat{u}_t,p_{t+1},1),\hat{u}_t-u_t\rangle\geq 0$$

or $p_{t+1}\circ D_2f(\hat{x}_t,\hat{u}_t)+\beta^tD_2\psi(\hat{x}_t,\hat{u}_t)=0$.

One can weaken the hypothesis of concavity of H_t with respect to x_t and u_t and replace it by the concavity of $\check{H}_t$ with respect to x_t as the following theorem shows. Let

$$\check{H}_t(x_t,p_{t+1},1)=\max_{u_t\in U_t}H_t(x_t,u_t,p_{t+1},1).$$

The maximum is attained since we shall assume that U_t is compact.

The following result of convex analysis [82] will be useful in the proof of the next theorem.

Lemma 3.11. *Let A be a convex subset of $\mathbb{R}^n$ and γ a real concave function defined on A. Let $\hat{z}$ be an interior point of A. Let ϕ be a real function defined on a ball $B(\hat{z},\delta)$ such that ϕ is differentiable at $\hat{z}$, $\phi(\hat{z})=\gamma(\hat{z})$ et $\phi(z)\leq\gamma(z)$, for all $z\in B(\hat{z},\delta)$.*

Then $\gamma(z)-\gamma(\hat{z})\leq D\phi(\hat{z}).(z-\hat{z})$.

Theorem 3.7. *Let U_t be closed for every t. Let $(\underline{\hat{x}},\underline{\hat{u}})\in\mathscr{X}\times\mathscr{U}\cap\mathrm{Adm}^e_\eta$. We assume that there exists $(p_t)_{t\in\mathbb{N}_*}\in\ell^1(\mathbb{N}_*,\mathbb{R}^{n*})$ and that the following conditions are fulfilled:*

(i) The mappings ψ and f are of class C^0 on $\Omega\times U$.
For all $u\in U$, the partial mappings $x\mapsto\psi(x,u)$ and $x\mapsto f(x,u)$ are of class C^1 on Ω. The mappings $D_1\psi$ and D_1f are of class C^0 on $\Omega\times U$.
(ii) and (iii) of the previous theorem.
(iv) For all $t\in\mathbb{N}$, the mapping $\check{H}_t$ is concave with respect to x_t, for all t.

Then $(\underline{\hat{x}},\underline{\hat{u}})$ is a solution of $(\mathscr{B}_{\mathbf{e}})$.

Proof. Let $(\underline{x},\underline{u})\in\mathscr{X}\times\mathscr{U}\cap\mathrm{Adm}^e_\eta$. By the definition of $\check{H}_t$ and noticing that $H_t(\hat{x}_t,\hat{u}_t,p_{t+1},1)=\check{H}_t(\hat{x}_t,p_{t+1},1)$, for all $t\in\mathbb{N}$, we have

$$\begin{aligned}
&\beta^t\psi(\hat{x}_t,\hat{u}_t) - \beta^t\psi(x_t,u_t)\\
&\quad = H_t(\hat{x}_t,\hat{u}_t,p_{t+1},1) - H_t(x_t,u_t,p_{t+1},1) - \langle p_{t+1}, f(\hat{x}_t,\hat{u}_t) - f(x_t,u_t)\rangle\\
&\quad \geq \check{H}_t(\hat{x}_t,p_{t+1},1) - \check{H}_t(x_t,p_{t+1},1) - \langle p_{t+1}, \hat{x}_{t+1} - x_{t+1}\rangle\\
&\quad = \check{H}_t(\hat{x}_t,p_{t+1},1) - \check{H}_t(x_t,p_{t+1},1) - \langle D_1H_{t+1}(\hat{x}_{t+1},\hat{u}_{t+1},p_{t+2},1), \hat{x}_{t+1} - x_{t+1}\rangle.
\end{aligned}$$

So using $\hat{x}_0 = x_0$, we obtain

$$\begin{aligned}
&\sum_{t=0}^{T}(\beta^t\psi(\hat{x}_t,\hat{u}_t) - \beta^t\psi(x_t,u_t))\\
&\quad \geq \sum_{t=0}^{T}(\check{H}_t(\hat{x}_t,p_{t+1},1) - \check{H}_t(x_t,p_{t+1},1) - \langle D_1H_t(\hat{x}_t,\hat{u}_t,p_{t+1},1), \hat{x}_t - x_t\rangle\,)\\
&\quad\quad -\langle p_{T+1}, \hat{x}_{T+1} - x_{T+1}\rangle.
\end{aligned}$$

The concavity of the maximized Hamiltonian with respect to x_t allows to use the previous lemma 3.11 to obtain

$$\check{H}_t(x_t,p_{t+1},1) - \check{H}_t(\hat{x}_t,p_{t+1},1) \leq \langle D_1H_t(\hat{x}_t,\hat{u}_t,p_{t+1},1), x_t - \hat{x}_t\rangle.$$

Hence $\sum_{t=0}^{T}(\beta^t\psi(\hat{x}_t,\hat{u}_t) - \beta^t\psi(x_t,u_t)) \geq \langle p_{T+1}, x_{T+1} - \hat{x}_{T+1}\rangle$.
Finally $J(\underline{\hat{x}},\underline{\hat{u}}) - J(\underline{x},\underline{u}) \geq 0$ follows as in the proof of Theorem 3.6. □

These theorems can be adapted to the unbounded case by adding an appropriate transversality condition:

Corollary 3.1. *Let* $(\underline{\hat{x}},\underline{\hat{u}}) \in \mathrm{Dom}_\eta^e(J)$ *(respectively* Adm_η^e*). If the hypotheses of Theorem 3.6 or Theorem 3.7 are satisfied except that* $(p_t)_{t\in\mathbb{N}_*} \in \ell^1(\mathbb{N}_*,\mathbb{R}^{n*})$ *is replaced by* $(p_t)_{t\in\mathbb{N}_*} \in (\mathbb{R}^{n*})^{\mathbb{N}_*}$ *and if the following hypothesis is also satisfied:*

(v) $\liminf_{T\to+\infty}\langle p_{T+1}, x_{T+1} - \hat{x}_{T+1}\rangle = 0,$

then $(\underline{\hat{x}},\underline{\hat{u}})$ *is a solution of* $(\mathscr{P}_{\mathbf{e}}^{\mathbf{s}})$ *(respectively* $(\mathscr{P}_{\mathbf{e}}^{\mathbf{o}})$*.)*

If $(\underline{\hat{x}},\underline{\hat{u}}) \in \mathrm{Adm}_\eta^e$ *with* $\limsup_{T\to+\infty}\langle p_{T+1}, x_{T+1} - \hat{x}_{T+1}\rangle = 0$*, we obtain that* $(\underline{\hat{x}},\underline{\hat{u}})$ *is a solution of* $(\mathscr{P}_{\mathbf{e}}^{\mathbf{w}})$.

Theorem 3.8. *Let* $(\underline{\hat{x}},\underline{\hat{u}}) \in \mathrm{int}\,\ell^\infty(\mathbb{N},\mathbb{R}^n)_+ \times \mathrm{int}\,\ell^\infty(\mathbb{N},\mathbb{R}^d)_+$ *be an admissible process. We assume that there exists* $(p_t)_{t\in\mathbb{N}_*} \in \ell^1(\mathbb{N}_*,\mathbb{R}^{n*})_+$ *such that the following conditions are satisfied:*

(i) The mappings ψ *and* f *are of class* C^1 *on* $\mathbb{R}_+^n \times \mathbb{R}_+^d$.
(ii) For all $t \in \mathbb{N}_*$, $p_t = p_{t+1} \circ D_1f_t(\hat{x}_t,\hat{u}_t) + \beta^t D_1\psi(\hat{x}_t,\hat{u}_t)$.
(iii) For all $t \in \mathbb{N}$, $p_{t+1} \circ D_2f(\hat{x}_t,\hat{u}_t) + \beta^t D_2\psi(\hat{x}_t,\hat{u}_t) = 0$.
(iv) For all $t \in \mathbb{N}$, $\langle p_{t+1}, f(\hat{x}_t,\hat{u}_t) - \hat{x}_{t+1}\rangle = 0$.
(v) For all $t \in \mathbb{N}$, *the mapping* H_t *is concave with respect to* (x_t,u_t).

Then $(\underline{\hat{x}},\underline{\hat{u}})$ *is a solution of* $(\mathscr{B}_{\mathrm{i}})$.

Proof. **A First Proof**

Recall that the adjoint equation (ii) is $p_t = D_1 H_t(\hat{x}_t, \hat{u}_t, p_{t+1}, 1)$ and the weak maximum principle (iii) is $D_2 H_t(\hat{x}_t, \hat{u}_t, p_{t+1}, 1) = 0$.

Let $(\underline{x}, \underline{u}) \in \ell^\infty(\mathbb{N}, \mathbb{R}^n) \times \ell^\infty(\mathbb{N}, \mathbb{R}^d) \cap \mathrm{Adm}^e_\eta$. Under hypothesis (iv) and since for all $t \in \mathbb{N}$, $x_{t+1} \leq f(x_t, u_t)$ and $p_{t+1} \geq 0$, we have for all $t \in \mathbb{N}$,

$$
\begin{aligned}
&\beta^t \psi(\hat{x}_t, \hat{u}_t) - \beta^t \psi(x_t, u_t) \\
&\quad = H_t(\hat{x}_t, \hat{u}_t, p_{t+1}, 1) - H_t(x_t, u_t, p_{t+1}, 1) - \langle p_{t+1}, f(\hat{x}_t, \hat{u}_t) - f(x_t, u_t) \rangle \\
&\quad \geq H_t(\hat{x}_t, \hat{u}_t, p_{t+1}, 1) - H_t(x_t, u_t, p_{t+1}, 1) - \langle p_{t+1}, \hat{x}_{t+1} - x_{t+1} \rangle, \\
&\quad = H_t(\hat{x}_t, \hat{u}_t, p_{t+1}, 1) - H_t(x_t, u_t, p_{t+1}, 1) - \langle D_1 H_{t+1}(\hat{x}_{t+1}, \hat{u}_{t+1}, p_{t+2}, 1), \hat{x}_{t+1} - \\
&\qquad\qquad x_{t+1} \rangle - \langle D_2 H_t(\hat{x}_t, \hat{u}_t, p_{t+1}, 1), \hat{u}_t - u_t \rangle + \langle D_2 H_t(\hat{x}_t, \hat{u}_t, p_{t+1}, 1), \hat{u}_t - u_t \rangle.
\end{aligned}
$$

Using $\hat{x}_0 = x_0$ we have

$$
\begin{aligned}
&\sum_{t=0}^{T} (\beta^t \psi(\hat{x}_t, \hat{u}_t) - \beta^t \psi(x_t, u_t)) \\
&\quad \geq \sum_{t=0}^{T} (H_t(\hat{x}_t, \hat{u}_t, p_{t+1}, 1) - H_t(x_t, u_t, p_{t+1}, 1) \\
&\quad -\langle D_1 H_t(\hat{x}_t, \hat{u}_t, p_{t+1}, 1), \hat{x}_t - x_t \rangle - \langle D_2 H_t(\hat{x}_t, \hat{u}_t, p_{t+1}, 1), \hat{u}_t - u_t \rangle) \\
&\quad -\langle p_{T+1}, \hat{x}_{T+1} - x_{T+1} \rangle + \sum_{t=0}^{T} \langle D_2 H_t(\hat{x}_t, \hat{u}_t, p_{t+1}, 1), \hat{u}_t - u_t \rangle.
\end{aligned}
$$

The concavity of H_t with respect to (x_t, u_t) for all t in $\mathbb{N}$ and hypothesis (iii) lead to $\sum_{t=0}^{T} (\beta^t \psi(\hat{x}_t, \hat{u}_t) - \beta^t \psi(x_t, u_t)) \geq \langle p_{T+1}, x_{T+1} - \hat{x}_{T+1} \rangle$. The end of the proof is similar to that of Theorem 3.6. □

A Second Proof

We introduce the reduced Lagrangian $\mathscr{R} : \ell^\infty(\mathbb{N}, \mathbb{R}^n) \times \ell^\infty(\mathbb{N}, \mathbb{R}^d) \to \mathbb{R}$ defined by

$$
\left.
\begin{aligned}
\mathscr{R}(\underline{x}, \underline{u}) := J(\underline{x}, \underline{u}) &+ \sum_{t=0}^{\infty} \langle p_{t+1}, f(x_t, u_t) - x_{t+1} \rangle \\
&-\langle D_1 \psi(\hat{x}_0, \hat{u}_0) + p_1 \circ D_1 f(\hat{x}_0, \hat{u}_0), x_0 - \eta \rangle.
\end{aligned}
\right\} \tag{3.17}
$$

Note that we can also write

$$
\left.
\begin{aligned}
\mathscr{R}(\underline{x}, \underline{u}) = &\sum_{t=0}^{\infty} H_t(x_t, u_t, 1, p_{t+1}) - \sum_{t=1}^{\infty} \langle p_t, x_t \rangle \\
&-\langle D_1 \psi(\hat{x}_0, \hat{u}_0) + p_1 \circ D_1 f(\hat{x}_0, \hat{u}_0), x_0 \rangle \\
&+\langle D_1 Psi(\hat{x}_0, \hat{u}_0) + p_1 \circ D_1 f(\hat{x}_0, \hat{u}_0), \eta \rangle.
\end{aligned}
\right\} \tag{3.18}
$$

First we split $\mathscr{R}$ in five functionals $\mathscr{R} = \mathscr{R}_1 + \mathscr{R}_2 + \mathscr{R}_3 + \mathscr{R}_4 + \mathscr{R}_5$ where

$$\left.\begin{aligned}
\mathscr{R}_1(\underline{x},\underline{u}) &= J(\underline{x},\underline{u})\\
\mathscr{R}_2(\underline{x},\underline{u}) &= \sum_{t=0}^{\infty}\langle p_{t+1}, f(x_t,u_t)\rangle\\
\mathscr{R}_3(\underline{x},\underline{u}) &= -\sum_{t=0}^{\infty}\langle p_{t+1}, x_{t+1}\rangle\\
\mathscr{R}_4(\underline{x},\underline{u}) &= -\langle D_1\psi(\hat{x}_0,\hat{u}_0) + p_1\circ D_1 f(\hat{x}_0,\hat{u}_0), x_0\rangle\\
\mathscr{R}_5(\underline{x},\underline{u}) &= \langle D_1\psi(\hat{x}_0,\hat{u}_0) + p_1\circ D_1 f(\hat{x}_0,\hat{u}_0), \eta\rangle
\end{aligned}\right\} \tag{3.19}$$

The functional $\mathscr{R}$ is of class C^1 as the sum of five functionals of class C^1 and we have the following formula:

$$\left.\begin{aligned}
D\mathscr{R}(\underline{\hat{x}},\underline{\hat{u}})(\underline{\delta x},\underline{\delta u}) &= \sum_{t=1}^{\infty}\langle \beta^t D_1\psi(\hat{x}_t,\hat{u}_t) + p_{t+1}\circ D_1 f(\hat{x}_t,\hat{u}_t) - p_t, \delta x_t\rangle\\
&\quad+\sum_{t=0}^{\infty}\langle \beta^t D_2\psi(\hat{x}_t,\hat{u}_t) + p_{t+1}\circ D_2 f(\hat{x}_t,\hat{u}_t, \delta u_t\rangle.
\end{aligned}\right\}$$

And then, using the adjoint equation (ii) and the weak maximum principle (ii) and (iii), we obtain

$$D\mathscr{R}(\underline{\hat{x}},\underline{\hat{u}}) = 0. \tag{3.20}$$

Now we want to prove that $\mathscr{R}$ is a concave functional. Using (3.18) we introduce the functional $\mathscr{R}_6 : \ell^\infty(\mathbb{N},\mathbb{R}^n)\times\ell^\infty(\mathbb{N},\mathbb{R}^d)\to\mathbb{R}$ defined as follows:

$$\mathscr{R}_6(\underline{x},\underline{u}) := \sum_{t=0}^{\infty} H_t(x_t,u_t,p_{t+1},1). \tag{3.21}$$

Note that we have $\mathscr{R} = \mathscr{R}_6 + \mathscr{R}_3 + \mathscr{R}_4 + \mathscr{R}_5$. $\mathscr{R}_5$ is constant; therefore it is concave. $\mathscr{R}_3$ and $\mathscr{R}_4$ are linear; therefore they are concave. Using the assumption (iv) we know that the function $h_t : (x,u)\mapsto H_t(x,u,p_{t+1},1)$, from $\mathbb{R}^n\times\mathbb{R}^d$ into $\mathbb{R}$, is concave. We introduce the mapping $\omega_t : \ell^\infty(\mathbb{N},\mathbb{R}^n)\times\ell^\infty(\mathbb{N},\mathbb{R}^d)\to\mathbb{R}^n\times\mathbb{R}^d$ by setting $\omega_t(\underline{x},\underline{u}) := (x_t,u_t)$. ω_t is linear and consequently $h_t\circ\omega_t$ is concave. Since a finite sum of concave functionals is concave, we obtain that $\sum_{t=0}^{T} h_t\circ\omega_t$ is concave for all $T\in\mathbb{N}$. Since a pointwise limit of concave functionals is concave, we obtain that $\mathscr{R}_6 = \lim_{T\to\infty}\sum_{t=0}^{T} h_t\circ\omega_t$ is concave. Consequently, since a finite sum of concave functionals is concave, we obtain

$$\mathscr{R} \text{ is concave.} \tag{3.22}$$

Since $\mathscr{R}$ is Fréchet differentiable at $(\underline{\hat{x}},\underline{\hat{u}})$ and since the Fréchet differentiability implies the Gâteaux differentiability, its subdifferential at $(\underline{\hat{x}},\underline{\hat{u}})$ is reduced to

$\{D\mathscr{R}(\underline{\hat{x}}, \underline{\hat{u}})\}$, [43] (Proposition 5.3, p. 23). Using the definition of the subdifferential, we have, for all $(\underline{x}, \underline{u}) \in \ell^\infty(\mathbb{N}, \mathbb{R}^n) \times \ell^\infty(\mathbb{N}, \mathbb{R}^d)$,

$$\mathscr{R}(\underline{x}, \underline{u}) \leq \mathscr{R}(\underline{\hat{x}}, \underline{\hat{u}}) + D\mathscr{R}(\underline{\hat{x}}, \underline{\hat{u}})(\underline{x} - \underline{\hat{x}}, \underline{u} - \underline{\hat{u}})$$

Then using (3.20) we obtain

$$\forall (\underline{x}, \underline{u}) \in \ell^\infty(\mathbb{N}, \mathbb{R}^n) \times \ell^\infty(\mathbb{N}, \mathbb{R}^d), \quad \mathscr{R}(\underline{x}, \underline{u}) \leq \mathscr{R}(\underline{\hat{x}}, \underline{\hat{u}}). \tag{3.23}$$

Using (iv) and $\hat{x}_0 = \eta$ we have

$$\mathscr{R}(\underline{\hat{x}}, \underline{\hat{u}}) = J(\underline{\hat{x}}, \underline{\hat{u}}). \tag{3.24}$$

Let $(\underline{x}, \underline{u})$ be an admissible process for $(\mathscr{P})$. Since $p_{t+1} \geq 0$ and $f(x_t, u_t) - x_{t+1} \geq 0$, we have $\langle p_{t+1}, f(x_t, u_t) - x_{t+1} \rangle \geq 0$, and since $x_0 = \eta$, we obtain

$$\mathscr{R}(\underline{x}, \underline{u}) \geq J(\underline{x}, \underline{u}). \tag{3.25}$$

Then using (3.23) and (3.24), we obtain, for all admissible process $(\underline{x}, \underline{u})$, that $J(\underline{\hat{x}}, \underline{\hat{u}}) \geq J(\underline{x}, \underline{u})$, and so $(\underline{\hat{x}}, \underline{\hat{u}})$ is a solution of $(\mathscr{B}_i)$. □

Corollary 3.2. *Let $(\underline{\hat{x}}, \underline{\hat{u}}) \in \mathrm{Dom}^i_\eta(J)$ (respectively Adm^i_η). If the hypotheses of Theorem 3.8 are satisfied except that $(p_t)_{t \in \mathbb{N}_*} \in \ell^1(\mathbb{N}_*, \mathbb{R}^{n*})_+$ is replaced by $(p_t)_{t \in \mathbb{N}_*} \in (\mathbb{R}^{n*}_+)^{\mathbb{N}_*}$ and if the following hypothesis is also satisfied: (v) $\liminf_{T \to +\infty} \langle p_{T+1}, x_{T+1} - \hat{x}_{T+1} \rangle = 0$, then $(\underline{\hat{x}}, \underline{\hat{u}})$ is a solution of $(\mathscr{P}^{\mathrm{s}}_{\mathrm{i}})$ (respectively $(\mathscr{P}^{\mathrm{o}}_{\mathrm{i}})$.) If $(\underline{\hat{x}}, \underline{\hat{u}}) \in \mathrm{Adm}^i_\eta$ with $\limsup_{T \to +\infty} \langle p_{T+1}, x_{T+1} - \hat{x}_{T+1} \rangle = 0$, we obtain that $(\underline{\hat{x}}, \underline{\hat{u}})$ is a solution of $(\mathscr{P}^{\mathrm{w}}_{\mathrm{i}})$.*

Theorem 3.9. *Let $(\underline{\hat{x}}, \underline{\hat{u}}) \in \mathrm{int}\, \ell^\infty(\mathbb{N}, \mathbb{R}^n)_+ \times \mathrm{int}\, \ell^\infty(\mathbb{N}, \mathbb{R}^d)_+$ be an admissible process. Under (H6) we assume that there exists $(p_t)_{t \in \mathbb{N}_*} \in \ell^1(\mathbb{N}_*, \mathbb{R}^{n*})_+$ such that the following conditions are satisfied:*

(i) The mappings ψ and f are of class C^2 on $\mathbb{R}^n \times \mathbb{R}^d$.
(ii), (iii), and (iv) of Theorem 3.8.
(v) For all $t \in \mathbb{N}$, there exists $c_t \in (0, \infty)$ such that, for all $\delta x_t \in \mathbb{R}^n$ and for all $\delta u_t \in \mathbb{R}^d$, $D_{11} H_t(\hat{x}_t, \hat{u}_t, p_{t+1}, 1)(\delta x_t, \delta x_t) + 2 D_{12} H_t(\hat{x}_t, \hat{u}_t, p_{t+1}, 1)(\delta x_t, \delta u_t) + D_{22} H_t(\hat{x}_t, \hat{u}_t, p_{t+1}, 1)(\delta u_t, \delta u_t) \leq -c_t(|\delta x_t|^2 + |\delta u_t|^2)$. And moreover we assume that $\sum_{t=0}^{\infty} c_t < \infty$.

Then there exists $r \in (0, +\infty)$ such that, for all admissible processes $(\underline{x}, \underline{u})$ satisfying $\|\underline{x} - \underline{\hat{x}}\| < r$ and $\|\underline{u} - \underline{\hat{u}}\| < r$, we have $J(\underline{\hat{x}}, \underline{\hat{u}}) \geq J(\underline{x}, \underline{u})$, i.e., $(\underline{\hat{x}}, \underline{\hat{u}})$ is a local solution of $(\mathscr{B}_{\mathrm{i}})$.

The problem is translated as a static optimization problem in the Banach space of sequences; hypothesis (v) guarantees local concavity: the second differential of the function $(x, u) \mapsto H_t(x, u, p_{t+1}, 1)$ is negative definite at $(\hat{x}_t, \hat{u}_t)$ for all $t \in \mathbb{N}$; it is just a local concavity condition. The additional condition on the c_t ensures that this property is conserved on our functionals defined on sequence spaces. And such a local property implies only a local conclusion.

Proof. Proceeding as in the proof of Lemma 3.9 we obtain that J is of class C^2. As it is shown in the proof of Theorem 3.8 we have $\mathscr{R} = \mathscr{R}_6 + \mathscr{R}_3 + \mathscr{R}_4 + \mathscr{R}_5$. Since $\mathscr{R}_3$ and $\mathscr{R}_4$ are linear, and $\mathscr{R}_5$ is constant, we have $D^2\mathscr{R} = D^2\mathscr{R}_6$. Using Theorem 4I.2 in Appendix in [17] we obtain that, for all $(\underline{\delta x}, \underline{\delta u}) \in \ell^\infty(\mathbb{N}, \mathbb{R}^n) \times \ell^\infty(\mathbb{N}, \mathbb{R}^d)$,

$$\begin{aligned} D^2\mathscr{R}(\underline{\hat{x}}, \underline{\hat{u}})((\underline{\delta x}, \underline{\delta u}), (\underline{\delta x}, \underline{\delta u})) &= D^2\mathscr{R}_6(\underline{\hat{x}}, \underline{\hat{u}})((\underline{\delta x}, \underline{\delta u}), (\underline{\delta x}, \underline{\delta u}))) \\ &= \sum_{t=0}^{\infty} D_{11}H_t(\hat{x}_t, \hat{u}_t, 1, p_{t+1})(\delta x_t, \delta x_t) \\ &\quad + 2D_{12}H_t(\hat{x}_t, \hat{u}_t, 1, p_{t+1})(\delta x_t, \delta u_t) \\ &\quad + D_{22}H_t(\hat{x}_t, \hat{u}_t, 1, p_{t+1})(\delta u_t, \delta u_t) \\ &\leq \sum_{t=0}^{\infty} (-c_t)(|\delta x_t|_\infty^2 + |\delta u_t|_\infty^2) \end{aligned}$$

which implies

$$D^2\mathscr{R}(\underline{\hat{x}}, \underline{\hat{u}})((\underline{\delta x}, \underline{\delta u}), (\underline{\delta x}, \underline{\delta u})) \leq -\left(\sum_{t=0}^{\infty} c_t\right) . (\|\underline{\delta x}\|_\infty^2 + \|\underline{\delta u}\|_\infty^2). \tag{3.26}$$

Using the continuity of $D^2\mathscr{R}$, there exists $r \in (0, \infty)$ such that when $(\underline{x}, \underline{u}) \in \ell^\infty(\mathbb{N}, \mathbb{R}^n)_+ \times \ell^\infty(\mathbb{N}, \mathbb{R}^d)_+$ satisfies $\|\underline{x} - \underline{\hat{x}}\|_\infty < r$ and $\|\underline{u} - \underline{\hat{u}}\|_\infty < r$, we have

$$D^2\mathscr{R}(\underline{x}, \underline{u})((\underline{\delta x}, \underline{\delta u}), (\underline{\delta x}, \underline{\delta u})) \leq -\frac{1}{2}\left(\sum_{t=0}^{\infty} c_t\right) . (\|\underline{\delta x}\|_\infty^2 + \|\underline{\delta u}\|_\infty^2).$$

And so, setting

$$\mathscr{N} := \{(\underline{x}, \underline{u}) \in \ell^\infty(\mathbb{N}, \mathbb{R}^n)_+ \times \ell^\infty(\mathbb{N}, \mathbb{R}^d)_+ : \|\underline{x} - \underline{\hat{x}}\|_\infty < r, \|\underline{u} - \underline{\hat{u}}\|_\infty < r\},$$

we obtain that $D^2\mathscr{R}(\underline{x}, \underline{u})$ is negative definite for all $(\underline{x}, \underline{u}) \in \mathscr{N}$, and therefore $\mathscr{R}$ is concave on $\mathscr{N}$. Now proceeding as in the end of the proof of Theorem 3.8, we obtain that $(\underline{\hat{x}}, \underline{\hat{u}})$ is a local maximizer of $(\mathscr{B}_i)$ on an open convex subset $\mathscr{N}$. □

Chapter 4
Related Topics

Stochastic Setting. For the Pontryagin principles of the discrete-time infinite-horizon stochastic optimal control, the pioneering work is that of G. Chow (see the reference in [10]). For the scalar case there exists such a principle in [10] which is based on a reduction to finite horizon and on a work in the finite-horizon setting due to Arkin and Evstigneev.

Perturbation, Regular Dependence. On discrete-time infinite-horizon calculus of variations problems, like the maximization of functionals as $\sum_{t=0}^{+\infty} \phi_t(x_t, x_{t+1}, \pi)$ (where π is a parameter), an interesting question is the dependence of the optimal solutions with respect to the parameters. Such a question is generally called the *sensitivity*. One possible way is to work on the Euler–Lagrange equation, $D_2\phi_{t-1}(x_{t-1}, x_t, \pi) + D_1\phi_t(x_t, x_{t+1}, \pi) = 0$. And then the study of the linearized equation of the Euler–Lagrange equation (an analogous equation to the Jacobi equation of the continuous time) allows to provide conditions to ensure a regular (continuous or differentiable) dependence. In this direction, there exists the famous Dominant Diagonal Blocs method that can be found in the paper of Araujo and Scheinkman cited in [17]. In this paper of Blot and Crettez, we find two original other methods to treat the linearized equation. It is useful to say that the study of this linearized equation needs (and so motivates) the study of certain classes of solutions of linear difference equations; see also [29].

Turnpike. For discrete-time infinite-horizon optimal control problems or calculus of variations problems, the main reference on the turnpike phenomenon is the book of Zaslavski [94]. The text of McKenzie [68] is also useful. In the paper of Blot and Crettez [18] we find a result of turnpike which uses a functional analytic setting around the space of the sequences which converge toward zero at infinity.

Continuous Time and Infinite Horizon. The unique and important book on this theory is the book of Carlson, Haurie and Leizarowitz [34]. The first result on Pontryagin principles in this setting is due to Halkin (see reference in [34]) which

J. Blot and N. Hayek, *Infinite-Horizon Optimal Control in the Discrete-Time Framework*, SpringerBriefs in Optimization, DOI 10.1007/978-1-4614-9038-8_4,

uses a method of reduction to finite horizon. Halkin also provides a counterexample to show that the usual condition of transversality of the finite horizon (i.e., the adjoint function is equal to zero at the final time) does not hold when the horizon is infinite. In the framework of the calculus of variations in infinite horizon, all the classical necessary conditions of optimality of the finite-horizon setting are established in the papers [27] (Blot and Michel) and [20] (Blot and Hayek); these works improve ancient works (due to Faedo and Cinquini) by using lighter assumptions. In [27], the authors use a method of reduction to finite horizon and a result of rounding off corners; in [20] the authors use the distributions of Schwartz. In presence of holonomic constraints, necessary conditions are established in [8]. Sufficient conditions and conjugate points are studied in (Blot and Hayek) ([21] and [22]). Again in the variational framework, problems in infinite horizon on bounded functions are studied in the papers of Blot and Cartigny [14] and [15], and in [19] by using functional analytic methods in Banach spaces.

Other Points. As it is mentioned about the continuous time, the transversality condition at infinity necessitates additional assumptions; in the continuous-time setting an example of such a condition is given in [27]. On the discrete-time setting, there exists a deep work of Michel on this transversality condition; it is [72]. In another work, [28], it is established that an infinite-horizon discrete-time problem with a quadratic criterion and a bilinear controlled dynamical system possesses a value function which is also quadratic.

Appendix A
Sequences

A.1 Limsup and Liminf

Let $(r_t)_{t\in\mathbb{N}}$ be a real sequence. Then

$$\limsup_{t\to+\infty} r_t := \lim_{T\to+\infty} (\sup_{t\geq T} r_t) = \inf_{T\in\mathbb{N}} (\sup_{t\geq T} r_t)$$

and

$$\liminf_{t\to+\infty} r_t := \lim_{T\to+\infty} (\inf_{t\geq T} r_t) = \sup_{T\in\mathbb{N}} (\inf_{t\geq T} r_t)$$

We denote by $\mathscr{S}(\mathbb{N},\mathbb{N})$ the set of all increasing functions from $\mathbb{N}$ into $\mathbb{N}$. Recall that, when $\sigma \in \mathscr{S}(\mathbb{N},\mathbb{N})$, we have $\sigma(t) \geq t$ for all $t \in \mathbb{N}$. A subsequence of the sequence $(r_t)_{t\in\mathbb{N}}$ is a sequence in the form $(r_{\sigma(t)})_{t\in\mathbb{N}}$ where $\sigma \in \mathscr{S}(\mathbb{N},\mathbb{N})$. We denote by $\mathrm{Adh}(r_t)_{t\in\mathbb{N}}$ the set of all $s \in [-\infty,+\infty]$ such that there exists $\sigma \in \mathscr{S}(\mathbb{N},\mathbb{N})$ for which the subsequence $(r_{\sigma(t)})_{t\in\mathbb{N}}$ converges to s. Then we have the following equalities:

$$\limsup_{t\to+\infty} r_t = \max \mathrm{Adh}(r_t)_{t\in\mathbb{N}} \quad \text{and} \quad \liminf_{t\to+\infty} r_t = \min \mathrm{Adh}(r_t)_{t\in\mathbb{N}}.$$

After the definitions and the characterizations, we give a list of the main properties of these notions.

1. The limit of $(r_t)_{t\in\mathbb{N}}$ exists in $[-\infty,+\infty]$ if and only if $\limsup_{t\to+\infty} r_t = \liminf_{t\to+\infty} r_t$.
   And then we have $\lim_{t\to+\infty} r_t = \limsup_{t\to+\infty} r_t = \liminf_{t\to+\infty} r_t$.
2. $\limsup_{t\to+\infty}(-r_t) = -\liminf_{t\to+\infty} r_t$.
3. $\limsup_{t\to+\infty}(r_t + s_t) \leq \limsup_{t\to+\infty} r_t + \limsup_{t\to+\infty} s_t$ when the second member is not indeterminate.

J. Blot and N. Hayek, *Infinite-Horizon Optimal Control in the Discrete-Time Framework*, SpringerBriefs in Optimization, DOI 10.1007/978-1-4614-9038-8,

4. $\liminf_{t\to+\infty}(r_t + s_t) \geq \liminf_{t\to+\infty} r_t + \liminf_{t\to+\infty} s_t$ when the second member is not indeterminate.
5. $\liminf_{t\to+\infty}(r_t + s_t) \leq \limsup_{t\to+\infty} r_t + \liminf_{t\to+\infty} s_t \leq \limsup_{t\to+\infty}(r_t + s_t)$ when the central term is not indeterminate.
6. When the limit of $(r_t)_{t\in\mathbb{N}}$ exists in $[-\infty, +\infty]$, we have

$$\limsup_{t\to+\infty}(r_t + s_t) = \lim_{t\to+\infty} r_t + \limsup_{t\to+\infty} s_t$$

and

$$\liminf_{t\to+\infty}(r_t + s_t) = \lim_{t\to+\infty} r_t + \liminf_{t\to+\infty} s_t.$$

A.2 A Diagonal Process of Cantor

The aim of this section is to prove the following result which is a way to express the diagonal process of Cantor. This is a result of compactness. It is also possible to prove it by using the theorem of Tychonov on the products of compact sets. On the diagonal process of Cantor, we can see [59]. We use the following notation: when $k \in \mathbb{N}$, $[k, +\infty)_{\mathbb{N}} := [k, +\infty) \cap \mathbb{N}$, and $\mathscr{S}([k, +\infty)_{\mathbb{N}} \to \mathbb{N})$ stands for the set of all strictly increasing functions from $[k, +\infty)_{\mathbb{N}}$ into $\mathbb{N}$.

Theorem A.1. *Let E be a finite-dimensional normed space. For all $t, T \in \mathbb{N}$ such that $t < T$, we consider $z_t^T \in E$. We assume that, for all $t \in \mathbb{N}$, the sequence $T \mapsto z_t^T$, from $[t+1, +\infty)_{\mathbb{N}}$ into E, is bounded. Then there exist a strictly increasing function $\delta : \mathbb{N}_* \to \mathbb{N}_*$ and a sequence $(w_t)_{t\in\mathbb{N}}$ in E such that $\lim_{T\to+\infty} z_t^{\delta(T)} = w_t$ for all $t \in \mathbb{N}$.*

Proof. First recall an important property of the elements of $\mathscr{S}([k, +\infty)_{\mathbb{N}} \to \mathbb{N})$.

$$\forall k \in \mathbb{N}, \forall \sigma \in \mathscr{S}([k, +\infty)_{\mathbb{N}} \to \mathbb{N}), \forall t \in [k, +\infty)_{\mathbb{N}}, \sigma(t) \geq t. \tag{A.1}$$

The proof of this property is easy by induction. Using the Bolzano–Weierstrass theorem, since $T \mapsto z_0^T$ is bounded and since $\dim E < +\infty$, there exist $\sigma_0 \in \mathscr{S}([1, +\infty)_{\mathbb{N}} \to \mathbb{N})$ and $w_0 \in E$ such that $\lim_{T\to+\infty} z_0^{\sigma_0(T)} = w_0$. Using the same argument, since $T \mapsto z_1^{\sigma_0(T)}$ is bounded, there exists $\sigma_1 \in \mathscr{S}([2, +\infty)_{\mathbb{N}} \to \mathbb{N})$ and $w_1 \in E$ such that $\lim_{T\to+\infty} z_1^{\sigma_0\circ\sigma_1(T)} = w_1$. Iterating these arguments, we obtain the following property.

$$\forall t \in \mathbb{N}, \exists \sigma \in \mathscr{S}([t + 1, +\infty)_{\mathbb{N}}), \exists w_t \in E, \lim_{T\to+\infty} z_t^{\sigma_0\circ\ldots\circ\sigma_t(T)} = w_t. \tag{A.2}$$

Now we define the "diagonal" function $\delta : \mathbb{N}_* \to \mathbb{N}_*$ by setting

$$\forall T \in \mathbb{N}_*, \ \ \delta(T) := \sigma_0 \circ \ldots \circ \sigma_T(T). \tag{A.3}$$

Then, using the increasingness of the functions and (A.1), we have

$$\begin{aligned}\delta(T+1) &= \sigma_0 \circ \ldots \circ \sigma_{T+1}(T+1)\\ &= \sigma_0 \circ \ldots \circ \sigma_T(\sigma_{T+1}(T+1))\\ &\geq \sigma_0 \circ \ldots \circ \sigma_T(T+1)\\ &> \sigma_0 \circ \ldots \circ \sigma_T(T) = \delta(T).\end{aligned}$$

And so we have proven

$$\delta \in \mathscr{S}(\mathbb{N}_*, \mathbb{N}). \tag{A.4}$$

Now we want to prove that $T \mapsto z_t^{\delta(T)}$ is a subsequence of $T \mapsto z_t^{\sigma_0 \circ \ldots \circ \sigma_t(T)}$. For all $t \in \mathbb{N}$, we define $\beta_t : [t+1, +\infty)_{\mathbb{N}} \to \mathbb{N}$ by setting

$$\forall T \in [t+1, +\infty)_{\mathbb{N}}, \ \ \beta_t(T) := \sigma_{t+1} \circ \ldots \circ \sigma_T(T). \tag{A.5}$$

For all $t \in \mathbb{N}$, for all $T \in [t+1, +\infty)_{\mathbb{N}}$, using (A.1) we have

$$\begin{aligned}\beta_t(T+1) &= \sigma_{t+1} \circ \ldots \circ \sigma_{T+1}(T+1) = \sigma_{t+1} \circ \ldots \circ \sigma_T(\sigma_{T+1}(T+1))\\ &\geq \sigma_{t+1} \circ \ldots \circ \sigma_T(T+1) > \sigma_{t+1} \circ \ldots \circ \sigma_T(T) = \beta_t(T).\end{aligned}$$

And so we have proven that $\beta_t \in \mathscr{S}([t+1, +\infty)_{\mathbb{N}}, \mathbb{N})$ and we have $\delta = (\sigma_0 \circ \ldots \circ \sigma_t) \circ \beta_t$, and then we can say that $T \mapsto z_t^{\delta(T)}$ is a subsequence of $T \mapsto z_t^{\sigma_0 \circ \ldots \circ \sigma_t(T)}$, and after (A.2) we obtain $\lim_{T \to +\infty} z_t^{\delta(T)} = w_t$ for all $t \in \mathbb{N}$. □

We can indicate another way to prove this result by using the theorem of Tychonov on the product of compacts (cf. [81]). Since $T \mapsto z_t^T$ is bounded and $\dim E < +\infty$, the closure $K_t := \mathrm{cl}\{z_t^T : T \in [t+1, +\infty)_{\mathbb{N}}\}$ is compact. K_t is also a metric space since E is normed. Using the theorem of Tychonov, $\mathfrak{K} := \prod_{t \in \mathbb{N}} K_t$ is a compact topological space when it is endowed with the cartesian product topology. We can also consider $\mathfrak{K}$ as a metric space with the metric $d((k_t)_{t\in\mathbb{N}}, (k_t')_{t\in\mathbb{N}}) := \sup_{t\in\mathbb{N}} \min\{\|k_t - k_t'\|, \frac{1}{t+1}\}$. The topology which is generated by this metric coincides with the cartesian product topology ([81], Théorème T2, VII, 1, 1, p. 61 and [31]). To have $T \mapsto z_t^T$ defined on all over $\mathbb{N}$ we can set $z_t^T := z_t^{t+1}$ when $T \in \{0, \ldots, t\}$. And so we have a sequence $T \mapsto (z_t^T)_{t\in\mathbb{N}}$ in $\mathfrak{K}$. Since $\mathfrak{K}$ is metric compact, there exist a strictly increasing function $\sigma : \mathbb{N}_* \to \mathbb{N}_*$ and $(w_t)_{t\in\mathbb{N}}$ such that $\lim_{T\to+\infty} d((z_t^T)_{t\in\mathbb{N}}, (w_t)_{t\in\mathbb{N}}) = 0$ which implies $\lim_{T\to+\infty} \|z_t^T - w_t\| = 0$.

A.3 Sequence Spaces

In this section we provide elements on sequence spaces, essentially on the space $\ell^{\infty}(\mathbb{N}, \mathbb{R}^k)$ of the bounded sequences in $\mathbb{R}^k$. First we define our notation and we recall some basic facts. Then we deal with the positive cone of $\ell^{\infty}(\mathbb{N}, \mathbb{R}^k)$, since we need one of its properties to apply a Karush–Kuhn–Tucker theorem in infinite-dimensional spaces. Finally we recall results on the dual space of $\ell^{\infty}(\mathbb{N}, \mathbb{R}^k)$ and we establish a proposition which is useful in the treatment of Pontryagin principles for bounded processes.

The main references that we use on the sequence spaces are Chap. 15 in the book of Aliprantis and Border [2] and Sect. 31 in the book of Köthe [60].

A.3.1 *Notation and Recall*

When $k \in \mathbb{N}_* := \mathbb{N} \setminus \{0\}$, on R^k we consider the norm

$$|v|_{\infty} := \max\{|v^j| : j \in \{1, \ldots, k\}\},$$

where $v = (v^1, \ldots, v^k)$. The natural norm on the dual $\mathbb{R}^{k*}$ is

$$|p|_* := \sup\{|\langle p, v\rangle| : v \in \mathbb{R}^k, |v|_{\infty} \leq 1\}$$

where $\langle p, v\rangle := p(v)$ is the duality bracket. Note that $|p|_* = \sum_{j=1}^{k} |p_j|$, where $p_j := \langle p, e_j\rangle$, $(e_j)_{1 \leq j \leq k}$ being the canonical basis of $\mathbb{R}^k$, $e^i_j := \delta^i_j$ (Kronecker symbol).

When $(E, |.|)$ is a finite-dimensional normed real vector space, we consider the following sets of sequences in E:

- $E^{\mathbb{N}}$ is the set of all sequences in E, and $\underline{x} = (x_t)_{t \in \mathbb{N}}$ denotes an element of $E^{\mathbb{N}}$.
- For all $p \in [1, +\infty)$, $\ell^p(\mathbb{N}, E) := \{\underline{x} \in E^{\mathbb{N}} : \sum_{t=0}^{+\infty} |x_t|^p < +\infty\}$. Endowed with the norm $\|\underline{x}\|_p := (\sum_{t=0}^{+\infty} |x_t|^p)^{1/p}$, it is a Banach space.
- $\ell^{\infty}(\mathbb{N}, E) := \{\underline{x} \in E^{\mathbb{N}} : \sup\{|x_t| : t \in \mathbb{N}\} < +\infty\}$. Endowed with the norm $\|\underline{x}\|_{\infty} := \sup\{|x_t| : t \in \mathbb{N}\}$, it is Banach space which is not reflexive.
- $c(\mathbb{N}, E) := \{\underline{x} \in E^{\mathbb{N}} : \lim_{t \to +\infty} x_t$ exists in $E\}$. It is a Banach subspace of $\ell^{\infty}(\mathbb{N}, E)$.
- $c_0(\mathbb{N}, E) := \{\underline{x} \in E^{\mathbb{N}} : \lim_{t \to +\infty} x_t = 0\}$. It is a Banach subspace of $\ell^{\infty}(\mathbb{N}, E)$.
- For all $p, q \in [1, +\infty)$ such that $p \leq q$, the following inclusions hold:

$$\ell^p(\mathbb{N}, E) \subset \ell^q(\mathbb{N}, E) \subset c_0(\mathbb{N}, E) \subset c(\mathbb{N}, E) \subset \ell^\infty(\mathbb{N}, E).$$

These notations are usual, and the proofs of the announced properties are given in the two abovementioned references.

A.3.2 About the Orders

$\mathbb{R}^k$ is endowed with its natural order

$$x \leq y \Longleftrightarrow (\forall j \in \{1, \ldots, k\},\ x^j \leq y^j).$$

We set $\mathbb{R}^k_+ := \{x \in \mathbb{R}^k : x \geq 0\}$; it is called the positive cone of $\mathbb{R}^k$.

On $\mathbb{R}^{k*}$ the order is defined by

$$p \leq q \Longleftrightarrow (\forall x \in \mathbb{R}^k_+,\ \langle p, x \rangle \leq \langle q, x \rangle).$$

When $p_i := \langle p, e_i \rangle$ and $q_i := \langle q, e_i \rangle$, we have

$$p \leq q \Longleftrightarrow (\forall i \in \{1, ..;, k\},\ p_i \leq q_i).$$

When $E = \mathbb{R}^k$ or $\mathbb{R}^{k*}$, on $E^{\mathbb{N}}$ we consider the following order:

$$\underline{x} \leq \underline{y} \Longleftrightarrow (\forall t \in \mathbb{N},\ x_t \leq y_t).$$

The considered order on each of the spaces $\ell^p(\mathbb{N}, E)$, when $p \in [1, +\infty]$, $c(\mathbb{N}, E)$ and $c_0(\mathbb{N}, E)$, is the restriction of the order of $E^{\mathbb{N}}$. We denote the positive cones by $\ell^p(\mathbb{N}, E)_+ := \{\underline{x} \in \ell^p(\mathbb{N}, E) :\ \underline{x} \geq \underline{0}\}$, where $\underline{0}$ is the constant sequence equal to zero. Similarly $c(\mathbb{N}, E)_+ := \{\underline{x} \in c(\mathbb{N}, E) :\ \underline{x} \geq \underline{0}\}$ and $c_0(\mathbb{N}, E)_+ := \{\underline{x} \in c_0(\mathbb{N}, E) :\ \underline{x} \geq \underline{0}\}$.

If we denote by $(E_+)^{\mathbb{N}}$ the set of all sequences in E_+, then we can define $\ell^p(\mathbb{N}, E_+) := \ell^p(\mathbb{N}, E) \cap (E_+)^{\mathbb{N}}$, $c(\mathbb{N}, E_+) := c(\mathbb{N}, E) \cap (E_+)^{\mathbb{N}}$ and $c_0(\mathbb{N}, E_+) := c_0(\mathbb{N}, E) \cap (E_+)^{\mathbb{N}}$.

Note that $\ell^p(\mathbb{N}, E)_+ = \ell^p(\mathbb{N}, E_+)$, $c(\mathbb{N}, E)_+ = c(\mathbb{N}, E_+)$ and $c_0(\mathbb{N}, E)_+ = c_0(\mathbb{N}, E_+)$.

Lemma A.1. *Let $\underline{x} \in \ell^\infty(\mathbb{N}, \mathbb{R}^k)_+$. Then the three following assertions are equivalent.*

(i) $\underline{x} \in \text{int}(\ell^\infty(\mathbb{N}, \mathbb{R}^k)_+)$.
(ii) There exists $a \in (0, +\infty)$ such that $x_t \geq a^{(k)}$.
(iii) For all $i \in \{1, \ldots, k\}$, $\inf\{x^i_t : t \in \mathbb{N}\} > 0$.

Proof. The equivalence between the two last assertions is easy: (ii) implies (iii) since $\inf\{x^i_t : t \in \mathbb{N}\} \geq a > 0$, and conversely it suffices to set

$$a := \min\{\inf\{x^i_t : t \in \mathbb{N}\} : i \in \{1, \ldots, k\}\} > 0.$$

When $\underline{x} \in \text{int}(\ell^\infty(\mathbb{N}, \mathbb{R}^k)_+)$ there exists $a \in (0, +\infty)$ such that the closed ball centered at $\underline{x}$ with a radius equal to a is included in $\ell^\infty(\mathbb{N}, \mathbb{R}^k)_+$. We denote by $\underline{a}$ the constant sequence equal to $a^{(k)} = (a, \dots, a)$ and therefore $\underline{x} - \underline{a}$ belongs to this ball (since $\|(\underline{x} - \underline{a} - \underline{x}\|_\infty = \|\underline{a}\|_\infty = a)$ which implies that $\underline{x} - \underline{a} \geq \underline{0}$, and consequently we have $x_t \geq a^{(k)}$ for all $t \in \mathbb{N}$; that proves $(i \Longrightarrow ii)$.

Conversely if (ii) holds, when $\underline{z} \in \ell^\infty(\mathbb{N}, \mathbb{R}^k)$ satisfies $\|\underline{z} - \underline{x}\|_\infty \leq a$, then, for all $t \in \mathbb{N}$, we have $|z_t - x_t|_\infty$ which implies, for all $i \in \{1, \dots, k\}$, $|z_t^i - x_t^i| \leq a$ and therefore we obtain $z_t^i \geq x_t^i + a \geq a > 0$. We have proven that the ball centered at $\underline{x}$ with a radius equal to a is included in $\ell^\infty(\mathbb{N}, \mathbb{R}^k)_+$ and consequently (i) holds. □

Proposition A.1. *The interior of $\ell^\infty(\mathbb{N}, \mathbb{R}^k)_+$ is nonempty.*

Proof. Using the previous lemma, we see that the sequence which is constant, equal to 1 at each time $t \in \mathbb{N}$, belongs to this interior. □

Lemma A.2. *Let $p \in [1, +\infty)$ and $\underline{x} \in \ell^p(\mathbb{N}, \mathbb{R}^k)_+$. Then, for all $\epsilon > 0$, there exists $\underline{z} \in \ell^p(\mathbb{N}, \mathbb{R}^k) \setminus \ell^p(\mathbb{N}, \mathbb{R}^k)_+$ such that $\|\underline{x} - \underline{z}\|_p < \epsilon$.*

Proof. We fix $\epsilon \in (0, +\infty)$. Since $\ell^p(\mathbb{N}, \mathbb{R}^k) \subset c_0(\mathbb{N}, \mathbb{R}^k)$, we know that there exists $t_\epsilon \in \mathbb{N}$ such that $|x_t| < \frac{\epsilon}{2}$ when $t \geq t_\epsilon$. We define $\underline{z} \in (\mathbb{R}^k)^{\mathbb{N}}$ by setting

$$z_t := \begin{cases} x_t & \text{if } t \neq t_\epsilon \\ -\frac{\epsilon}{2} 1^{(n)} & \text{if } t = t_\epsilon. \end{cases}$$

Using properties of real series, since $(|z_t|^p)_{t > t_\epsilon} = (|x_t|^p)_{t > t_\epsilon}$ and since $\underline{x} \in \ell^p(\mathbb{N}, \mathbb{R}^k)$, we have $\underline{z} \in \ell^p(\mathbb{N}, \mathbb{R}^k)$. Since $z_{t_\epsilon} \notin \mathbb{R}^k_+$, we have $\underline{z} \notin \ell^p(\mathbb{N}, \mathbb{R}^k)_+$. We calculate

$$\|\underline{z} - \underline{x}\|_p = (0 + |z_{t_\epsilon} - x_{t_\epsilon}|^p)^{1/p} \leq |z_{t_\epsilon}| + |x_{t_\epsilon}| < \frac{\epsilon}{2} + \frac{\epsilon}{2} = \epsilon.$$

□

Proposition A.2. *When $p \in [1, +\infty)$ then the interior of $\ell^p(\mathbb{N}, \mathbb{R}^k)_+$ is empty.*

Proof. Using the previous lemma, when $\underline{x} \in \ell^p(\mathbb{N}, \mathbb{R}^k)_+$, there does not exist any ball centered at $\underline{x}$ which is included in $\ell^p(\mathbb{N}, \mathbb{R}^k)_+$, and this proves that no point of $\ell^p(\mathbb{N}, \mathbb{R}^k)_+$ has $\ell^p(\mathbb{N}, \mathbb{R}^k)_+$ for a neighborhood. □

This result is known; it is given in [40] (p. 219) and also in [62] (p. 18) without proof. We have decided to provide a proof since we don't know a reference where an explicit proof is given.

A.3.3 The Dual Space of $\ell^\infty(\mathbb{N}, \mathbb{R}^k)$

In [2] (p. 507) the following space is defined.

Definition A.1. $\ell^1_d(\mathbb{N}, \mathbb{R})$ is the set of all linear functionals $\theta \in \ell^\infty(\mathbb{N}, \mathbb{R})^*$ such that there exists $\xi \in \mathbb{R}$ satisfying $\langle \theta, \underline{x} \rangle = \xi . \lim_{t \to +\infty} x_t$ for all $\underline{x} \in c(\mathbb{N}, \mathbb{R})$. Its elements are called the singular functionals of $\ell^\infty(\mathbb{N}, \mathbb{R})^*$.

In [2] (p. 507) the following result is established.

Theorem A.2. $\ell^\infty(\mathbb{N},\mathbb{R})^* = \ell^1(\mathbb{N},\mathbb{R}) \oplus \ell^1_d(\mathbb{N},\mathbb{R})$.

The meaning of this equality is the following: for all $\Lambda \in \ell^\infty(\mathbb{N},\mathbb{R})^*$ there exists a unique $(\underline{q},\theta) \in \ell^1(\mathbb{N},\mathbb{R}) \times \ell^1_d(\mathbb{N},\mathbb{R})$ such that $\langle \Lambda, \underline{r}\rangle = \langle \underline{r}, \underline{q}\rangle + \langle \theta, \underline{r}\rangle$ for all $\underline{r} \in \ell^\infty(\mathbb{N},\mathbb{R})$.

Now we extend this space and the previous description to sequences in $\mathbb{R}^k$.

Definition A.2. $\ell^1_d(\mathbb{N},\mathbb{R}^k)$ is the set of all linear functionals $\theta \in \ell^\infty(\mathbb{N},\mathbb{R}^k)^*$ such that there exists $\zeta \in \mathbb{R}^{k*}$ satisfying $\langle \theta, \underline{x}\rangle = \langle \zeta, \lim_{t\to+\infty} x_t\rangle$ for all $\underline{x} \in c(\mathbb{N},\mathbb{R}^k)$. Its elements are called the singular functionals of $\ell^\infty(\mathbb{N},\mathbb{R}^k)^*$.

Proposition A.3. $\ell^\infty(\mathbb{N},\mathbb{R}^k)^* = \ell^1(\mathbb{N},\mathbb{R}^{k*}) \oplus \ell^1_d(\mathbb{N},\mathbb{R}^k)$.

Proof. Let $\Lambda \in \ell^\infty(\mathbb{N},\mathbb{R}^k)^*$. When $\underline{x} \in \ell^\infty(\mathbb{N},\mathbb{R}^k)$, we can identify it with $(\underline{x}^1,\dots,\underline{x}^k) \in \ell^\infty(\mathbb{N},\mathbb{R})^k$. And we can write $\langle \Lambda, \underline{x}\rangle = \sum_{i=1}^k \langle \Lambda_i, \underline{x}^i\rangle$ where

$$\langle \Lambda_i, \underline{x}^i\rangle = \langle \Lambda, (\underline{0},\dots,\underline{0},\underline{x}^i,\underline{0},\dots,\underline{0})\rangle.$$

Note that $\Lambda_i \in \ell^\infty(\mathbb{N},\mathbb{R})^*$ and then, using Theorem A.2, we know that there exist $\underline{q}^i \in \ell^1(\mathbb{N},\mathbb{R})$ and $\theta^i \in \ell^1_d(\mathbb{N},\mathbb{R})$ such that $\langle \Lambda_i, \underline{r}\rangle = \langle \underline{r}, \underline{q}^i\rangle + \langle \theta^i, \underline{r}\rangle$ for all $\underline{r} \in \ell^\infty(\mathbb{N},\mathbb{R})$.

Denoting by $(e_i^*)_{1\le i\le k}$ the dual basis of the canonical basis of $\mathbb{R}^k$, we set $q_t := \sum_{i=1}^k q_t^i e_i^* \in \mathbb{R}^{k*}$. Since $|q_t|_* = \sum_{i=1}^k |q_t^i|$ we obtain that $\underline{q} = (q_t)_{t\in\mathbb{N}} \in \ell^1(\mathbb{N},\mathbb{R}^{k*})$.

We set $\langle \theta, \underline{x}\rangle := \sum_{i=1}^k \langle \theta^i, \underline{x}^i\rangle$. We see that θ is a linear functional from $\ell^\infty(\mathbb{N},\mathbb{R}^k)$ into $\mathbb{R}$. Since the projections $\pi_i : \ell^\infty(\mathbb{N},\mathbb{R}^k) \to \ell^\infty(\mathbb{N},\mathbb{R})$, $\pi_i(\underline{x}) := \underline{x}^i$, are continuous, $\theta = \sum_{i=1}^k \theta^i \circ \pi_i$ is continuous as a finite sum of compositions of continuous functions. And so we obtain $\theta \in \ell^\infty(\mathbb{N},\mathbb{R}^k)^*$. When $\underline{x} \in c(\mathbb{N},\mathbb{R}^k)$ we have $\underline{x}^i \in c(\mathbb{N},\mathbb{R})$, and since $\theta^i \in \ell^1_d(\mathbb{N},\mathbb{R})$ there exists $\xi_i \in \mathbb{R}$ such that $\langle \theta^i, \underline{x}^i\rangle = \xi_i . \lim_{t\to+\infty} x^i{}_t$. We set $\xi := \sum_{i=1}^k \xi_i e_i^* \in \mathbb{R}^{k*}$, and then we have $\langle \theta, \underline{x}\rangle = \sum_{i=1}^k \langle \theta^i, \underline{x}^i\rangle = \sum_{i=1}^k \xi_i . \lim_{t\to+\infty} x_t^i = \langle \xi, \lim_{t\to+\infty} x_t\rangle$. and so $\theta \in \ell^1_d(\mathbb{N},\mathbb{R}^k)$. The existence is proven.

To justify the uniqueness we assume that there exist $\underline{p} \in \ell^1(\mathbb{N},\mathbb{R}^{k*})$ and $\rho \in \ell^1_d(\mathbb{N},\mathbb{R}^k)$ such that $\langle \underline{x}, \underline{q}\rangle + \langle \theta, \underline{x}\rangle = \langle \underline{x}, \underline{p}\rangle + \langle \rho, \underline{x}\rangle$ for all $\underline{x} \in \ell^\infty(\mathbb{N},\mathbb{R}^k)$. When

$\underline{x} \in c_0(\mathbb{N}, \mathbb{R}^k)$, this equality becomes $\langle \underline{x}, \underline{q} \rangle = \langle \underline{x}, \underline{p} \rangle$, and since $\ell^1(\mathbb{N}, \mathbb{R}^{k*})$ is the dual space of $c_0(\mathbb{N}, \mathbb{R}^k)$, we obtain $\underline{q} = \underline{p}$, from which we deduce $\theta = \rho$. □

Theorem A.3. *Let $\Lambda \in \ell^\infty(\mathbb{N}, \mathbb{R}^k)^*$; $\Lambda = \underline{q} + \theta$ with $\underline{q} \in \ell^1(\mathbb{N}, \mathbb{R}^{k*})$ and $\theta \in \ell^1_d(\mathbb{N}, \mathbb{R}^k)$. The following assertions hold.*

(a) $(\Lambda \geq 0) \Longleftrightarrow (\underline{q} \geq \underline{0} \text{ and } \theta \geq 0)$.
(b) We assume that $\Lambda \geq 0$. Let $\underline{w} \in \ell^\infty(\mathbb{N}, \mathbb{R}^k)_+$ such that $\langle \Lambda, \underline{w} \rangle = 0$. Then we have $\langle q_t, w_t \rangle = 0$ for all $t \in \mathbb{N}$.

Proof. **(a)** The implication ($\Longleftarrow$) is easy. Now we prove the converse. When $\underline{x} \in c_0(\mathbb{N}, \mathbb{R}^k)_+$ then $\langle \theta, \underline{x} \rangle = 0$ and consequently we have $\langle \underline{x}, \underline{q} \rangle = \langle \Lambda, \underline{w} \rangle \geq 0$. We fix $t \in \mathbb{N}$ and we consider $\underline{x}$ defined by $x_t := e_i$ (canonical basis) and $x_s := 0$ when $s \neq t$. Then we have $\underline{x} \in c_0(\mathbb{N}, \mathbb{R}^k)$ and $0 \leq \langle \underline{x}, \underline{q} \rangle = \langle q_t, e_i \rangle$. Since this last inequality holds for all $i \in \{1, \ldots, k\}$, we obtain that $q_t \geq 0$, and so $\underline{q} \geq \underline{0}$. Let $\underline{x} \in \ell^\infty(\mathbb{N}, \mathbb{R}^k)_+$. For all $T \in \mathbb{N}$ we define $\underline{x}^{[T]}$ by setting

$$x_t^{[T]} := \begin{cases} 0 \text{ if } t \leq T \\ x_t \text{ if } t > T. \end{cases}$$

Then we have $\underline{x}^{[T]} \in \ell^\infty(\mathbb{N}, \mathbb{R}^k)_+$ and we have

$$x_t - x_t^{[T]} := \begin{cases} x_t \text{ if } t \leq T \\ 0 \text{ if } t > T. \end{cases}$$

And so $\underline{x} - \underline{x}^{[T]} \in c_0(\mathbb{N}, \mathbb{R}^k)_+$ for all $T \in \mathbb{N}$, and consequently we have $\langle \theta, \underline{x} - \underline{x}^{[T]} \rangle = \langle \xi, 0 \rangle = 0$ which implies $\langle \theta, \underline{x} \rangle = \langle \theta, \underline{x}^{[T]} \rangle$ for all $T \in \mathbb{N}$. Using $\Lambda \geq 0$ we obtain

$$0 \leq \langle \Lambda, \underline{x}^{[T]} \rangle = \langle \underline{q}, \underline{x}^{[T]} \rangle + \langle \theta, \underline{x}^{[T]} \rangle = \sum_{t=T+1}^{+\infty} \langle q_t, x_t \rangle + \langle \theta, \underline{x}^{[T]} \rangle.$$

The convergence of the series $\sum_{t=0}^{+\infty} \langle q_t, x_t \rangle$ in $\mathbb{R}$ implies that $\lim_{T \to +\infty} \sum_{t=T+1}^{+\infty} \langle q_t, x_t \rangle = 0$. Therefore from the previous inequalities we deduce $0 \leq 0 + \langle \theta, \underline{x} \rangle$, and so we have proven that $\theta \geq 0$.

(b) Using (a), we have

$$0 = \langle \underline{w}, \underline{q} \rangle + \langle \theta, \underline{w} \rangle \geq \langle \underline{w}, \underline{q} \rangle \geq 0$$

since $\langle \theta, \underline{w} \rangle \geq 0$, and then we obtain $\langle \underline{w}, \underline{q} \rangle = 0$. Since $0 = \sum_{t=0}^{+\infty} \langle q_t, w_t \rangle$ with $\langle q_t, w_t \rangle \geq 0$ for all $t \in \mathbb{N}$, we obtain the announced conclusion. □

The assertion (a) is established in Lemma 4.3 in [26]. The assertion (b) is included into the second step of the proof of Theorem 3.1 in [26].

Appendix B
Static Optimization

In this appendix we want to provide results on necessary conditions of optimality for finite-dimensional static optimization problems. These necessary conditions are often given under assumptions of continuous differentiability (or of strict differentiability), or under conditions of convexity (or concavity); cf. [1, 46, 47, 49, 52, 74, 75, 88, 92].

In a first time, we provide a multiplier rule which was first established by Halkin in 1974. To prove his result, Halkin used an implicit function theorem in an only Fréchet differentiable setting instead of the classical continuously differentiable setting. We provide a proof in the spirit of the initial proof of Halkin in Sect. B.2. In such a proof there is a point which is not completely clear for us; this is why we prefer to present, in Sect. B.1, a proof of Michel, of 1989, which does not use an implicit function theorem and which appears completely clear for us.

Besides, there exist several generalizations of the usual differential calculus, in order to work with non Fréchet differentiable functions; the subdifferential of convex (or concave) functions is a kind of model for such a generalization. Among all these generalizations we have chosen to use the Clarke calculus and Sect. B.3 is devoted to this theory of Clarke who had established a multiplier rule for locally Lipschitzian functions. Finally in Sect. B.4, we present Karush–Kuhn–Tucker theorems in Banach spaces that will be useful in the study of the bounded case in Chap. 3.

B.1 A Theorem of Halkin and a Proof of Michel

We consider two nonnegative integer numbers n_I and n_E, a nonempty open subset Ω in $\mathbb{R}^n$, and functions $g^0, g^1, \ldots, g^{n_I}, h^1, \ldots, h^{n_E}$ from Ω into $\mathbb{R}$. With these elements we formulate the following maximization problem:

J. Blot and N. Hayek, *Infinite-Horizon Optimal Control in the Discrete-Time Framework*, SpringerBriefs in Optimization, DOI 10.1007/978-1-4614-9038-8,

$$(\mathcal{M}) \begin{cases} \text{Maximize } g^0(z) \\ \quad \text{when} \quad \forall \alpha \in \{1, \ldots, n_I\}, \; g^\alpha(z) \geq 0 \\ \qquad\qquad\quad \forall \beta \in \{1, \ldots, n_E\}, \; h^\beta(z) = 0. \end{cases}$$

The conditions $g^\alpha(z) \geq 0$ are called the inequality constraints, the conditions $h^\beta(z) = 0$ are called the equality constraints, and f is called the criterion. A point $z \in \Omega$ which satisfies all the inequality constraints and all the equality constraints is called *admissible* for $(\mathcal{M})$. Following Michel [72] we use the following vocabulary.

Definition B.1. The function $\mathcal{L} : \Omega \times \mathbb{R}^{n_I} \times \mathbb{R}^{n_E} \to \mathbb{R}$ defined by

$$\mathcal{L}(z, \lambda_1, \ldots, \lambda_{n_I}, \mu_1, \ldots, \mu_{n_E}) := g^0(z) + \sum_{\alpha=1}^{n_I} \lambda_\alpha g^\alpha(z) + \sum_{\beta=1}^{n_E} \mu_\beta h^\beta(z)$$

is called the Lagrangian of $(\mathcal{M})$.

The function $\mathcal{G} : \Omega \times \mathbb{R} \times \mathbb{R}^{n_I} \times \mathbb{R}^{n_E} \to \mathbb{R}$ defined by

$$\mathcal{G}(z, \lambda_0, \lambda_1, \ldots, \lambda_{n_I}, \mu_1, \ldots, \mu_{n_E}) := \sum_{\alpha=0}^{n_I} \lambda_\alpha g^\alpha(z) + \sum_{\beta=1}^{n_E} \mu_\beta h^\beta(z)$$

is called the generalized Lagrangian of $(\mathcal{M})$.

Note that the difference between $\mathcal{G}$ and $\mathcal{L}$ is the presence of a scalar λ_0 associated to the criterion.

The following theorem is established in the paper of Halkin [48] (published in 1974). In [71] (published in 1989) Michel provides a proof which is different from this one of Halkin.

Theorem B.1. *Let z_* be a solution of $(\mathcal{M})$. We assume that the functions g^0, g^1, $\ldots$, g^{n_I}, h^1,$\ldots$, h^{n_E} are continuous on a neighborhood of z_* and that they are Fréchet differentiable at z_*. Then there exist real numbers λ_0, λ_1, $\ldots$, λ_{n_I}, μ_1, $\ldots$, μ_{n_E} which satisfy the following conditions:*

(i) $\lambda_0, \lambda_1, \ldots, \lambda_{n_I}, \mu_1, \ldots, \mu_{n_E}$ are not simultaneously equal to zero.
(ii) For all $\alpha \in \{0, \ldots, n_I\}$, $\lambda_\alpha \geq 0$.
(iii) For all $\alpha \in \{1, \ldots, n_I\}$, $\lambda_\alpha g^\alpha(z_*) = 0$.
(iv) $D_1\mathcal{G}(z_*, \lambda_0, \lambda_1, \ldots, \lambda_{n_I}, \mu_1, \ldots, \mu_{n_E}) = 0$, where D_1 denotes the partial differential with respect to the first variable z.

The real numbers of the conclusion of the theorem are called the *multipliers* associated to z_*. λ_0 is called the multiplier associated to the criterion; when $\alpha \in \{1, \ldots, n_I\}$, λ_α is called the multiplier associated to the inequality constraint $g^\alpha(z) \geq 0$.; and when $\beta \in \{1, \ldots, n_E\}$, μ_β is called the multiplier associated to the equality constraint $h^\beta(z) = 0$. About the conclusion (i), it is easy to see that when all the multipliers are zero then all the conclusions hold even if z_* is not a solution of the problem. The conclusion (iii) is called the slackness condition; it means that when $g^\alpha(z_*) > 0$ then the associated multiplier λ_α is zero and consequently we can delete it. The conclusion (iv) can be translated as follows:

$$\sum_{\alpha=0}^{n_I} \lambda_\alpha Dg^\alpha(z_*) + \sum_{\beta=1}^{n_E} \mu_\beta Dh^\beta(z_*) = 0.$$

Note that when $(\lambda_0, \ldots, \mu_{n_E})$ satisfies (i–iv), for all $r \in (0, +\infty)$, the new list $(r.\lambda_0, \ldots, r.\mu_{nE})$ also satisfies (i–iv); it is a property of cones. Consequently it is possible to normalize a list $(\lambda_0, \ldots, \mu_{n_E})$ which satisfies (i–iv): choosing a norm $\|.\|$ on $\mathbb{R}^{1+n_I+n_E}$, we can choose a suitable list such that $\|(\lambda_0, \ldots, \mu_{n_E})\| = 1$. Also note that the set of all lists $(\lambda_0, \ldots, \mu_{n_E})$ which satisfy (i–iv) is a convex subset of $\mathbb{R}^{1+n_I+n_E}$.

A more classical theorem ensures the conclusions (i–iv) by assuming that the functions g^α and h^β are continuously differentiable; e.g., see [1, 88].

The proof of Halkin in [48] uses (among other arguments) an implicit function theorem for differentiable functions; the classical implicit function theorem uses continuously differentiable functions, [1]. In our knowledge, the proof of Michel exists only in [71] which is a book for teaching, written in French, and out of print today. It is why it seems useful to provide a variation of the proof of Michel; all the main ideas are due to Michel, and it is only our presentation which slightly differs. This proof uses two important tools that we recall now.

B.1.1 Two Tools

The first tool is the fixed-point theorem of Brouwer.

Theorem B.2. *Let K be a convex compact subset of $\mathbb{R}^n$ and $F : K \to K$ be a continuous function. Then there exists $\hat{x} \in K$ such that $f(\hat{x}) = \hat{x}$.*

There exist several proofs of this famous theorem that use simplicial topology or homology theory or Stokes formula or degree theory. We indicate only two references which contain elementary proofs, where *elementary* means without tools of the algebraic topology: they are the proof of Dunford and Schwartz in [41] and the proof of Milnor in [5, 73]. These elementary proofs are done in the case where the convex compact subset K is the unit closed ball. In [57], Kantorovitch and Akilov prove that when a convex compact subset has a nonempty interior, then it is homeomorphic to the unit closed ball (using the Minkowski functional). To avoid the assumption on the interior, it suffices to work in the relative topology, i.e., that of the affine hull of the convex subset, as it is currently done for many questions in the book of Rockafellar [79].

The second tool is a theorem on the existence of supporting hyperplane.

Theorem B.3. *Let C be a nonempty convex subset of $\mathbb{R}^n$ and $z \in \mathrm{bd}C$ (the topological boundary of C). Then there exist $\Lambda \in \mathscr{L}(\mathbb{R}^n, \mathbb{R})$, $\Lambda \neq 0$, and $c \in \mathbb{R}$ such that $\langle \Lambda, z \rangle = c$ and $\langle \Lambda, x \rangle \leq c$ for all $x \in C$.*

This result is a corollary of the Hahn–Banach theorem. It is proven in [89] and in [79] (Corollary 11.6.2).

B.1.2 A First Simplification

We introduce $A_* := \{\alpha \in \{1, \ldots, n_I\} : g^\alpha(z_*) = 0\}$ and the new maximization problem:

$$(\mathscr{M}1) \begin{cases} \text{Maximize } g^0(z) \\ \quad \text{when} \quad \forall \alpha \in A_*, \ \ g^\alpha(z) \geq 0 \\ \qquad\qquad\ \forall \beta \in \{1, \ldots, n_E\}, \ \ h^\beta(z) = 0. \end{cases}$$

If we have λ_0, $(\lambda_\alpha)_{\alpha \in A_*}$ and $(\mu_\beta)_{1 \leq \beta \leq n_E}$ which satisfy the following conditions:

(i') λ_0, $(\lambda_\alpha)_{\alpha \in A_*}$ and $(\mu_\beta)_{1 \leq \beta \leq n_E}$ are not simultaneously equal to zero
(ii') For all $\alpha \in \{0\} \cup A_*$, $\lambda_\alpha \geq 0$
(iv') $\displaystyle\sum_{\alpha \in \{0\} \cup A_*} \lambda_\alpha D g^\alpha(z_*) + \sum_{\beta=1}^{n_E} \mu_\beta h^\beta(z_*) = 0$

then by taking $\lambda_\alpha := 0$ when $\alpha \in \{1, \ldots, n_I\} \setminus A_*$, we obtain λ_0, $(\lambda_\alpha)_{1 \leq \alpha \leq n_I}$, $(\mu_\beta)_{1 \leq \beta \leq n_E}$ which satisfy the conclusions (i–iv) of Theorem B.1. And so it suffices to prove the theorem when $g^\alpha(z_*) = 0$ for all $\alpha \in \{1, \ldots, n_I\}$.

B.1.3 A Transformation of the Problem into a Separation Problem

We introduce the function $\Phi : \Omega \to \mathbb{R}^{1+n_I+n_E}$ by setting

$$\Phi(z) := (g^0(z) - g^0(z_*), g^1(z), \ldots, g^{n_I}(z), h^1(z), \ldots, h^{n_E}(z)). \tag{B.1}$$

We know that Φ is differentiable at z_* since all its coordinates are differentiable, and for all $\zeta \in \mathbb{R}^n$ we have

$$D\Phi(z_*).\zeta := (Dg^0(z_*).\zeta, Dg^1(z_*).\zeta, \ldots, Dg^{n_I}(z_*).\zeta, Dh^1(z_*).\zeta, \ldots, Dh^{n_E}(z_*).\zeta). \tag{B.2}$$

Note that $D\Phi(z_*) \in \mathscr{L}(\mathbb{R}^n, \mathbb{R}^{1+n_I+n_E})$.
The range of $D\Phi(z_*)$ is denoted by $\mathrm{Im}(D\Phi(z_*))$.

Lemma B.1. *$(\lambda_0, \lambda_1, \ldots, \lambda_{n_I}, \mu_1, \ldots, \mu_{n_E}) \in \mathbb{R}^{1+n_I+n_E}$ satisfies (i–iv) if and only if the functional $\Lambda \in \mathscr{L}(\mathbb{R}^{1+n_I+n_E}, \mathbb{R})$ defined by $\Lambda(v, w) := \sum_{\alpha=0}^{n_I} \lambda_\alpha v^\alpha + \sum_{\beta=1}^{n_E} \mu_\beta w^\beta$ is nonzero and satisfies $\Lambda(v, w) \leq 0$ for all $(v, w) \in \mathrm{Im}(D\Phi(z_*)) + (\mathbb{R}_-^{1+n_I} \times \{0\})$.*

Proof. If $(\lambda_0, \lambda_1, \ldots, \lambda_{n_I}, \mu_1, \ldots, \mu_{n_E}) \in \mathbb{R}^{1+n_I+n_E}$ satisfies (i–iv) then $\Lambda \neq 0$ follows from (i). When $(v, w) \in \mathrm{Im}(D\Phi(z_*)) + (\mathbb{R}_-^{1+n_I} \times \{0\})$, then there exist $\zeta \in \mathbb{R}^n$ and $\xi \in \mathbb{R}_-^{1+n_I}$ such that $(v, w) = D\Phi(z_*).\zeta + (\xi, 0)$, i.e.,

$$\begin{cases} \forall \alpha \in \{0, \ldots, n_I\},\ v^\alpha = Dg^\alpha(z_*).\zeta + \xi^\alpha \\ \forall \beta \in \{1, \ldots, n_E\},\ w^\beta = Dh^\beta(z_*).\zeta, \end{cases}$$

and then we have, using (iv) and (ii), the following relations:

$$\begin{aligned} \Lambda(v, w) &= \sum_{\alpha=0}^{n_I} \lambda_\alpha (Dg^\alpha(z_*).\zeta + \xi^\alpha) + \sum_{\beta=1}^{n_E} \mu_\beta Dh^\beta(z_*).\zeta \\ &= (\sum_{\alpha=0}^{n_I} \lambda_\alpha Dg^\alpha(z_*) + \sum_{\beta=1}^{n_E} \mu_\beta Dh^\beta(z_*)).\zeta + \sum_{\alpha=0}^{n_I} \lambda_\alpha \xi^\alpha \\ &= 0 + \sum_{\alpha=0}^{n_I} \lambda_\alpha \xi^\alpha \\ &\leq 0. \end{aligned}$$

Conversely we assume that $\Lambda(v, w) \leq 0$ for all $(v, w) \in \mathrm{Im}(D\Phi(z_*)) + (\mathbb{R}_-^{1+n_I} \times \{0\})$. Therefore, for all $\zeta \in \mathbb{R}^n$ and for all $\xi \in \mathbb{R}^{1+n_I}$, we have $\Lambda(D\Phi(z_*).\zeta + (\xi, 0)) \leq 0$, i.e.,

$$\sum_{\alpha=0}^{n_I} \lambda_\alpha (Dg^\alpha(z_*).\zeta + \xi^\alpha) + \sum_{\beta=1}^{n_E} \mu_\beta Dh^\beta(z_*).\zeta \leq 0. \tag{B.3}$$

Taking $\xi = 0$ in (B.3) we obtain, for all $\zeta \in \mathbb{R}^n$,

$$\sum_{\alpha=0}^{n_I} \lambda_\alpha Dg^\alpha(z_*).\zeta + \sum_{\beta=1}^{n_E} \mu_\beta Dh^\beta(z_*).\zeta \leq 0$$

and since $-\zeta \in \mathbb{R}^n$ we also obtain

$$-\sum_{\alpha=0}^{n_I} \lambda_\alpha Dg^\alpha(z_*).\zeta - \sum_{\beta=1}^{n_E} \mu_\beta Dh^\beta(z_*).\zeta \leq 0$$

which implies (iv).

Note that (iii) results from $g^\alpha(z_*) = 0$ for all $\alpha \in \{1,\dots,n_I\}$, and that the conclusion (i) results from $\Lambda \neq 0$. Taking $\zeta = 0$ in (B.3) we obtain, for all $\xi \in \mathbb{R}_-^{1+n_I}$, $\sum_{\alpha=0}^{n_I} \lambda_\alpha \xi^\alpha \leq 0$, which implies (ii). □

Setting $C := \mathrm{Im}(D\Phi(z_*)) + (\mathbb{R}_-^{1+n_I} \times \{0\})$ and using the previous lemma, we see that, if $0 \in \mathrm{bd}(C)$, Λ is a supporting hyperplane to the convex set C at 0. And using Theorem B.3 if $0 \in \mathrm{bd}(C)$, the existence of a supporting hyperplane is guaranteed, which implies that the existence of $(\lambda_0, \lambda_1, \dots, \lambda_{n_I}, \mu_1, \dots, \mu_{n_E}) \in \mathbb{R}^{1+n_I+n_E}$ which satisfies (i–iv) is ensured, and the theorem will be proven.

B.1.4 Last Step

It remains to prove that $0 \in \mathrm{bd}(C)$. To realize that, we proceed by contradiction, and we assume that $0 \notin \mathrm{bd}(C)$. Since $0 \in C$, which implies that $0 \in \mathrm{int}(C)$ (the topological interior of C). Since C is a neighborhood of 0, there exists $r \in (0, +\infty)$ such that the closed ball centered at 0 with a radius equal to r is included into C. Let $(e_1, \dots, e_{n_E})$ be the canonical basis of $\mathbb{R}^{n_E}$, and let $v_0 := (r, \dots, r) \in \mathbb{R}^{1+n_I}$. We have $(v_0, re_\beta) \in C$, for all $\beta \in \{1,\dots,n_E\}$, since its norm is equal to r. Using the definition of C, for all $\beta \in \{1,\dots,n_E\}$, there exist $\zeta_\beta \in \mathbb{R}^n$ and $\xi_\beta \in \mathbb{R}_-^{1+n_I}$ such that $(v_0, re_\beta) = D\Phi(z_*).\zeta_\beta + (\xi_\beta, 0)$. Similarly for all $\beta \in \{1,\dots,n_E\}$, there exist $\zeta'_\beta \in \mathbb{R}^n$ and $\xi'_\beta \in \mathbb{R}_+^{1+n_I}$ such that $(v_0, -re_\beta) = D\Phi(z_*).\zeta'_\beta + (\xi'_\beta, 0)$. We define $B := \mathrm{co}(\{\zeta_\beta : \beta \in \{1,\dots,n_E\}\} \cup \{\zeta'_\beta : \beta \in \{1,\dots,n_E\}\})$, where co means convex hull. Then B is a convex compact subset of $\mathbb{R}^n$, as the convex hull of a finite set after a theorem of Mazur (([41] p. 416) since a nonempty finite set is compact. We set $\sigma := \sup\{\|b\| : b \in B\}$; $\sigma \in (0, +\infty)$ since B is not reduced to 0.

The Fréchet differentiability of the function Φ given in (B.1) at z_* means

$$\forall \epsilon > 0, \exists \delta_\epsilon > 0, \forall \zeta \in \mathbb{R}^n, \|\zeta\| \leq \delta_\epsilon \Longrightarrow \|\Phi(z_*+\zeta) - \Phi(z_*) - D\Phi(z_*).\zeta\| \leq \epsilon \|\zeta\|. \tag{B.4}$$

Now we fix ϵ such that $0 < \epsilon < \frac{r}{2\sigma n_E}$ and we fix a such that $0 < a < \frac{\delta_\epsilon}{\sigma}$. Then when $\zeta \in B$, we have $\|a\zeta\| = a\|\zeta\| \leq a\sigma \leq \delta_\epsilon$ and by using (B.4) and $\Phi(z_*) = 0$ we obtain

$$\|\Phi(z_* + a\zeta) - D\Phi(z_*).(a\zeta)\| \leq \epsilon \|a\zeta\| \leq \frac{r}{2\sigma n_E} a\sigma = \frac{r}{2n_E} a,$$

and translating this inequality on the coordinates of Φ we obtain: there exists $a \in (0, +\infty)$ such that, for all $\zeta \in B$, the following relations hold.

$$|g^0(z_* + a\zeta) - g^0(z_*) - Dg^0(z_*).(a\zeta)| \leq \frac{r}{2n_E} a \tag{B.5}$$

$$\forall \alpha \in \{1, \dots, n_I\}, \quad |g^\alpha(z_* + a\zeta) - Dg^\alpha(z_*).(a\zeta)| \leq \frac{r}{2n_E} a \tag{B.6}$$

$$\forall \beta \in \{1, \dots, n_E\}, \quad |h^\beta(z_* + a\zeta) - Dh^\beta(z_*).(a\zeta)| \leq \frac{r}{2n_E} a. \tag{B.7}$$

Now we define, for all $\beta \in \{1, \dots, n_E\}$, the function $\omega_\beta : B \to \mathbb{R}$ by setting

$$\left.\begin{aligned} \omega_\beta(\zeta) &:= \tfrac{1}{a} h^\beta(z_* + a\zeta) - Dh^\beta(z_*).\zeta \\ \omega(\zeta) &:= (\omega_1(\zeta), \dots, \omega_{n_E}(\zeta)). \end{aligned}\right\} \tag{B.8}$$

We define the function $F : B \to \mathbb{R}^n$ by setting, for all $\zeta \in B$,

$$F(\zeta) := \sum_{\beta=0}^{n_E} (\frac{1}{2n_E} - \frac{1}{2r}\omega_\beta(\zeta))\zeta_\beta + \sum_{\beta=0}^{n_E} (\frac{1}{2n_E} + \frac{1}{2r}\omega_\beta(\zeta))\zeta'_\beta. \tag{B.9}$$

where ζ_β and ζ'_β are those of the definition of B.

We arbitrarily fix $\zeta \in B$. Note that, for all $\beta \in \{1, \dots, n_E\}$, following (B.7), we have $|\omega_\beta(\zeta)| \leq \frac{r}{n_E}$ which implies $\frac{1}{2n_E} - \frac{1}{2r}\omega_\beta(\zeta) \geq \frac{1}{2n_E} - \frac{1}{2r}\frac{r}{n_E} = 0$ and $\frac{1}{2n_E} + \frac{1}{2r}\omega_\beta(\zeta) \geq \frac{1}{2n_E} - \frac{1}{2r}\frac{r}{n_E} = 0$, and moreover

$$\begin{aligned} &\sum_{\beta=0}^{n_E} (\frac{1}{2n_E} - \frac{1}{2r}\omega_\beta(\zeta)) + \sum_{\beta=0}^{n_E} (\frac{1}{2n_E} + \frac{1}{2r}\omega_\beta(\zeta)) \\ &= n_E \tfrac{1}{2n_E} - \tfrac{1}{2r} \sum_{\beta=0}^{n_E} \omega_\beta(\zeta) + n_E \frac{1}{2n_E} + \frac{1}{2r} \sum_{\beta=0}^{n_E} \omega_\beta(\zeta) \\ &= \tfrac{1}{2} + \tfrac{1}{2} = 1, \end{aligned}$$

and so $F(\zeta)$ is a convex combination of the ζ_β and of the ζ'_β that ensures that $F(\zeta) \in B$.

And so we have defined a function $F : B \to B$, where B is convex and compact. Using the definition (B.8) and the continuity of h^β and the continuity of the Fréchet differential $Dh^\beta(z_*)$ (as a linear function on a finite-dimensional space) we can assert that ω_β is continuous, and consequently we deduce that F is continuous. Now we can use the fixed-point theorem of Brouwer (Theorem B.2) and we can assert that there exists $\hat{\zeta} \in B$ such that $F(\hat{\zeta}) = \hat{\zeta}$.

Now we verify that $z_* + a\hat{\zeta}$ is admissible for $(\mathscr{M})$. We set $h := (h^1, \dots, h^{n_E})$, and so $Dh(z_*).\zeta_\beta = re_\beta$ and $Dh(z_*).\zeta'_\beta = -re_\beta$. Then, when $\beta \in \{1, \dots, n_E\}$, we have

$$\begin{aligned}
Dh(z_*).\hat{\zeta} &= Dh(z_*).F(\hat{\zeta}) \\
&= \sum_{\beta=0}^{n_E}(\frac{1}{2n_E} - \frac{1}{2r}\omega_\beta(\hat{\zeta}))Dh(z_*).\zeta_\beta + \sum_{\beta=0}^{n_E}(\frac{1}{2n_E} + \frac{1}{2r}\omega_\beta(\hat{\zeta}))Dh(z_*).\zeta'_\beta \\
&= \sum_{\beta=0}^{n_E}(\frac{1}{2n_E} - \frac{1}{2r}\omega_\beta(\hat{\zeta}))(re_\beta) + \sum_{\beta=0}^{n_E}(\frac{1}{2n_E} + \frac{1}{2r}\omega_\beta(\hat{\zeta}))(-re_\beta) \\
&= \sum_{\beta=0}^{n_E}[\frac{1}{2n_E}re_\beta - \frac{1}{2r}\omega_\beta(\hat{\zeta})(re_\beta) - \frac{1}{2n_E}re_\beta - \frac{1}{2r}\omega_\beta(\hat{\zeta})(re_\beta)] \\
&= -\sum_{\beta=0}^{n_E}\omega_{\hat{\beta}}(\zeta)e_\beta = -\omega(\hat{\zeta}).
\end{aligned}$$

And then $Dh(z_*).\hat{\zeta} = -\frac{1}{a}h(z_* + a\hat{\zeta}) + Dh(z_*).\hat{\zeta}$ which implies $\frac{1}{a}h(z_* + a\hat{\zeta}) = 0$ and consequently $h(z_* + a\hat{\zeta}) = 0$ which implies

$$\forall \beta \in \{1, \ldots, n_E\}, \quad h^\beta(z_* + a\hat{\zeta}) = 0. \tag{B.10}$$

Now we fix $\alpha \in \{0, \ldots, n_I\}$. Following the definition of ζ_β, ξ_β, ζ'_β, ξ'_β, we have $r = Dg^\alpha(z_*).\zeta_\beta + \xi^\alpha_\beta \leq Dg^\alpha(z_*).\zeta_\beta$ since $\xi^\alpha_\beta \leq 0$ and $r = Dg^\alpha(z_*).\zeta'_\beta + \xi'^\alpha_\beta \leq Dg^\alpha(z_*).\zeta'_\beta$ with $\xi_\beta \leq 0$ since $\xi'^\alpha_\beta \leq 0$. Since $\hat{\zeta}$ is a convex combination of the ζ_β and of the ζ'_β, we have $\hat{\zeta} = \sum_{\beta=1}^{n_E} c_\beta\zeta_\beta + \sum_{\beta=1}^{n_E} c'_\beta\zeta'_\beta$ with $c_\beta \geq 0$, $c'_\beta \geq 0$ and $\sum_{\beta=1}^{n_E} c_\beta + \sum_{\beta=1}^{n_E} c'_\beta = 1$. Then

$$\begin{aligned}
Dg^\alpha(z_*).\hat{\zeta} &= \sum_{\beta=1}^{n_E} c_\beta Dg^\alpha(z_*).\zeta_\beta + \sum_{\beta=1}^{n_E} c'_\beta Dg^\alpha(z_*).\zeta'_\beta \\
&\geq \sum_{\beta=1}^{n_E} c_\beta r + \sum_{\beta=1}^{n_E} c'_\beta r \geq r.
\end{aligned}$$

And so we have proven the following relations:

$$\forall \alpha \in \{0, \ldots, n_E\}, \quad Dg^\alpha(z_*).\hat{\zeta} \geq r. \tag{B.11}$$

Using (B.6) and (B.11) we obtain, for all $\alpha \in \{1, \ldots, n_E\}$,

$$g^\alpha(z_* + a\hat{\zeta}) - Dg^\alpha(z_*).(a\hat{\zeta}) \geq -\frac{r}{2n_E}a$$

which implies

$$g^{\alpha}(z_* + a\hat{\xi}) \geq Dg^{\alpha}(z_*).(a\hat{\xi}) - \frac{r}{2n_E}a \geq ra - \frac{r}{2n_E}a \geq 0.$$

And so, using (B.10) and the previous inequalities, we have proven that $z_* + a\hat{\xi}$ is admissible for $(\mathscr{M})$.

Now using (B.5) and (B.11) we obtain

$$g^0(z_* + a\hat{\xi}) - g^0(z_*) - Dg^0(z_*).(a\hat{\xi}) \geq -\frac{r}{2n_E}a$$

which implies

$$g^0(z_* + a\hat{\xi}) - g^0(z_*) \geq Dg^0(z_*).(a\hat{\xi}) - \frac{r}{2n_E}a \geq ra - \frac{r}{2n_E}a > 0$$

that is a contradiction. And so the theorem is proven.

Remark B.1. In the previous proof it can seem necessary to have $n_E \geq 1$. We will see in the next section that the case without equality constraints is easier.

B.2 Another Proof of Halkin's Theorem

The proof of Halkin in [48] is based on an implicit function theorem which uses only the Fréchet differentiability instead of the continuous differentiability as it is usual in the differential calculus, e.g., in [1,64]. Now we recall this implicit function theorem of Halkin (Theorem E in [48] p. 230).

Theorem B.4. *Let X and Y be two normed spaces, let Z be a finite-dimensional Euclidean space, let ϕ be a mapping from $X \times Y$ into Z, and let A be a linear continuous mapping from X into Z and B be a linear continuous mapping from Y onto Z such that*

(i) ϕ is continuous on a neighborhood of $(\hat{x}, \hat{y})$.
(ii) (A, B) is the differential of ϕ at $(\hat{x}, \hat{y})$.

Then there exists a neighborhood U of $\hat{x}$, a mapping ψ from U into Y, and a linear continuous mapping C from X into Y such that

(α) $\psi(\hat{x}) = \hat{y}$.
(β) $\phi(x, \psi(x)) = \phi(\hat{x}, \hat{y})$.
(γ) C is the differential of ψ at $\hat{x}$.
(δ) $A + BC = 0$.

Remark B.2. Using the partial differentials, $A = D_1\phi(\hat{x}, \hat{y})$, $B = D_2\phi(\hat{x}, \hat{y})$, when $D_2\phi(\hat{x}, \hat{y})$ is invertible the condition of onto is fulfilled, and the conclusion (δ) is simply the consequence of the chain rule: $D_1\phi(\hat{x}, \hat{y}) + D_2\phi(\hat{x}, \hat{y}) \circ D\psi(\hat{x})$, from which we obtain the same formula than these one of the classical implicit function theorem:

$$D\psi(\hat{x}) = -(D_2\phi(\hat{x},\hat{y}))^{-1} \circ D_1\phi(\hat{x},\hat{y}).$$

It is useful to say that the two proofs that Halkin gives in [48] both use the fixed-point theorem of Brouwer.

Now we can give a sketch of a proof of Theorem B.1 in the spirit of the proof of Halkin.

B.2.1 First Step

We use the implicit function theorem to erase the equality constraints. When the differential $Dh(z_*)$ is onto splitting the space $\mathbb{R}^n = W \oplus S$ where $W := \mathrm{Ker}(Dh(z_*))$ and S is a supplementary subspace of $\mathrm{Ker}(Dh(z_*))$ in $\mathbb{R}^n$, the partial differential $D_2h(z_*)$ becomes invertible and we can use Theorem B.1 to ensure the existence of a neighborhood U of w_*, where $z_* = (w_*, s_*)$, and of a mapping $\psi : U \to S$ which is differentiable at w_* and such that, for all $w \in U$, and for all $\beta \in \{1,\dots,n_E\}$, $h^\beta(w,\psi(w)) = 0$. We define $\Gamma^\alpha(w) := g^\alpha(w,\psi(w))$ for all $\alpha \in \{0,\dots,n_I\}$. Therefore w_* is a solution of the following maximization problem:

$$(\mathcal{M}')\begin{cases} \text{Maximize } \Gamma^0(w) \\ \qquad \text{when } \forall\alpha \in \{1,\dots,n_I\}, \quad \Gamma^\alpha(w) \geq 0. \end{cases}$$

Using the same remark as of the previous section, we can assume that $\Gamma^\alpha(w_*) = 0$ for all $\alpha \in \{1,\dots,n_I\}$, and we can also assume that $\Gamma^0(w_*) = 0$ since we do not modify a maximization by adding a constant to the criterion.

B.2.2 A Transformation of the Problem into a Separation Problem

As in the previous section, we transform the problem of the existence of the multipliers into a problem of existence of a supporting hyperplane.

Lemma B.2. *Let* $(\nu_0,\dots,\nu_{n_I}) \in \mathbb{R}^{1+n_I}$*. Then* $(\nu_0,\dots,\nu_{n_I})$ *satisfies the following conditions:*

(a) $(\nu_0,\dots,\nu_{n_I}) \neq (0,\dots,0)$,
(b) $\forall\alpha \in \{1,\dots,n_I\}, \quad \nu_\alpha \geq 0$,
(c) $\sum_{\alpha=0}^{n_I} \nu_\alpha \, D\Gamma^\alpha(w_*) = 0$,

if and only if the linear function $\Lambda \in \mathscr{L}(\mathbb{R}^{1+n_I}, \mathbb{R})$ *defined by* $\langle \Lambda, y \rangle := \sum_{\alpha=0}^{n_I} \nu_\alpha y^\alpha$ *satisfies* $\langle \Lambda, y \rangle \leq 0$ *for all* $y \in \text{Im}(D\Gamma(w_*)) + \mathbb{R}_-^{1+n_I}$, *where* $\Gamma := (\Gamma^0, \ldots, \Gamma^{n_I})$.

Proof. If $(\nu_0, \ldots, \nu_{n_I})$ satisfies (a, b, c) then, for all $y \in \text{Im}(D\Gamma(w_*)) + \mathbb{R}_-^{1+n_I}$, there exist $\zeta \in W$ and $\xi \in \mathbb{R}_-^{1+n_I}$ such that $y = D\Gamma(w_*).\zeta + \xi$, and then we have:

$$\begin{aligned}\langle \Lambda, y \rangle &= \sum_{\alpha=0}^{n_I} \nu_\alpha (D\Gamma^\alpha(w_*).\zeta + \xi^\alpha) \\ &= \sum_{\alpha=0}^{n_I} \nu_\alpha D\Gamma^\alpha(w_*).\zeta + \sum_{\alpha=0}^{n_I} \nu_\alpha \xi^\alpha \\ &= 0 + \sum_{\alpha=0}^{n_I} \nu_\alpha \xi^\alpha \quad (\text{by}(c)) \\ &\leq 0\end{aligned}$$

after (b) by using $\xi^\alpha \leq 0$.

Conversely if Λ is nonpositive on $\text{Im}(D\Gamma(w_*)) + \mathbb{R}_-^{1+n_I}$, then for all $\zeta \in W$ and for all $\xi \in \mathbb{R}_-^{1+n_I}$, we have

$$0 \geq \langle \sum_{\alpha=0}^{n_I} \nu_\alpha D\Gamma^\alpha(w_*), \zeta \rangle + \sum_{\alpha=0}^{n_I} \nu_\alpha \xi^\alpha.$$

Taking $\xi = 0$ in the previous inequality we obtain $\langle \sum_{\alpha=0}^{n_I} \nu_\alpha D\Gamma^\alpha(w_*), \zeta \rangle \leq 0$ for all $\zeta \in W$, and since $-\zeta$ also belongs to W, we obtain (c). On the other hand, taking $\zeta = 0$ we obtain $\sum_{\alpha=0}^{n_I} \nu_\alpha \xi^\alpha \leq 0$ for all $\xi \in \mathbb{R}_-^{1+n_I}$ which implies (b). □

B.2.3 Multipliers for the Simplified Problem

And so to use Theorem B.3 to obtain $(\nu_0, \ldots, \nu_{n_I})$ which satisfies (a, b, c), it suffices to prove that 0 does not belong to the interior of $\text{Im}(D\Gamma(w_*)) + \mathbb{R}_-^{1+n_I}$. To prove that, we proceed by contradiction, we assume that the closed ball centered at 0 with a radius $r > 0$ is contained in $\text{Im}(D\Gamma(w_*)) + \mathbb{R}_-^{1+n_I}$. Then setting $v_0 := (r, \ldots, r) \in \mathbb{R}^{1+n_I}$ we can say that there exist $\zeta_0 \in W$ and $\xi_0 \in \mathbb{R}^{1+n_I}$ such that $v_0 = D\Gamma(w_*)\zeta_0 + \xi_0$, i.e., $r = D\Gamma^\alpha(w_*).\zeta_0 + \xi_0^\alpha$ for all $\alpha \in \{0, \ldots, n_I\}$. Since $\xi_0^\alpha \leq 0$ we obtain $D\Gamma(w_*)\zeta_0 \geq r$ for all $\alpha \in \{0, \ldots, n_I\}$. Using a similar argument to this one of (B.5) and (B.6) we obtain

$$|\Gamma^\alpha(w_* + a\zeta_0) - D\Gamma^\alpha(w_*).(a\zeta)| \leq \frac{ar}{2}$$

which implies

$$\Gamma^\alpha(w_* + a\zeta_0) \geq D\Gamma^\alpha(w_*).(a\zeta) - \frac{ar}{2} \geq ar - \frac{ar}{2} = \frac{ar}{2} > 0$$

which implies that $w_* + a\zeta_0$ is admissible for $(\mathscr{M}')$ and that $\Gamma^0(w_* + a\zeta_0) > 0 = \Gamma^0(w_*)$ that is a contradiction.

B.2.4 Last Step

Now we have obtained $(\nu_0, \ldots, \nu_{n_I})$ which satisfies (a, b, c). Using Remark B.2 we know that

$$D\psi(w_*) = -(D_2h(z_*))^{-1} \circ D_1h(z_*). \tag{B.12}$$

Using the chain rule we obtain, for all $\alpha \in \{0, \ldots, n_I\}$,

$$D\Gamma^\alpha(w_*) = D_1g^\alpha(z_*) + D_2g^\alpha(z_*) \circ D\psi(w_*). \tag{B.13}$$

From (c) we obtain

$$\begin{aligned}
0 &= \sum_{\alpha=0}^{n_I} \nu_\alpha D\Gamma^\alpha(w_*) \\
&= \sum_{\alpha=0}^{n_I} \nu_\alpha (D_1g^\alpha(z_*) + D_2g^\alpha(z_*) \circ D\psi(w_*)) \\
&= D_1(\sum_{\alpha=0}^{n_I} \nu_\alpha g^\alpha)(z_*) + D_2(\sum_{\alpha=0}^{n_I} \nu_\alpha g^\alpha)(z_*) \circ D\psi(w_*) \\
&= D_1(\sum_{\alpha=0}^{n_I} \nu_\alpha g^\alpha)(z_*) + D_2(\sum_{\alpha=0}^{n_I} \nu_\alpha g^\alpha)(z_*) \circ (-(D_2h(z_*))^{-1} \circ D_1h(z_*)).
\end{aligned}$$

We set

$$M := D_2(\sum_{\alpha=0}^{n_I} \nu_\alpha g^\alpha)(z_*) \circ (D_2h(z_*))^{-1}. \tag{B.14}$$

And then from the previous calculations we obtain

$$D_1(\sum_{\alpha=0}^{n_I} \nu_\alpha g^\alpha)(z_*) = M \circ D_1h(z_*)).$$

It is clear that the following equality holds

$$D_2(\sum_{\alpha=0}^{n_I} \nu_\alpha g^\alpha)(z_*) = M \circ D_2h(z_*)).$$

From these two last equalities we obtain

$$D(\sum_{\alpha=0}^{n_I} \nu_\alpha g^\alpha)(z_*) = M \circ Dh(z_*)). \tag{B.15}$$

Since $M \in \mathscr{L}(\mathbb{R}^{n_E}, \mathbb{R})$, in the canonical basis of $\mathbb{R}^{n_E}$, it is represented by the matrix with one row: $[m_1 \ldots m_{n_E}]$. Then the equality (B.5) becomes

$$D(\sum_{\alpha=0}^{n_I} \nu_\alpha g^\alpha)(z_*) = \sum_{\beta=1}^{n_E} m_\beta Dh^\beta(z_*),$$

which implies

$$\sum_{\alpha=0}^{n_I} \nu_\alpha Dg^\alpha(z_*) + \sum_{\beta=1}^{n_E}(-m_\beta)Dh^\beta(z_*) = 0. \tag{B.16}$$

And so, setting $\lambda_\alpha := \nu_\alpha$ for all $\alpha \in \{0, \ldots, n_I\}$ and $\mu_\beta := -m_\beta$ for all $\beta \in \{1, \ldots, n_E\}$, we obtain the wanted result and the proof is finished.

Remark B.3. In this last proof we have assumed the condition: $Dh(z_*)$ is onto, as Halkin in his paper. To obtain a complete proof it is necessary to show how to avoid this additional assumption. Note that this question does not arise in the proof of Michel.

B.3 A Theorem of Clarke

B.3.1 Strict Differentiability

First we consider two finite-dimensional normed vector spaces E and F, a nonempty subset A of E, a point $a \in A$, and a function $\phi : A \to F$. $E^* := \mathscr{L}(E, \mathbb{R})$ which denote the dual space of E.

Definition B.2. We assume that A is a neighborhood of a. ϕ is said to be strictly differentiable at a when there exists $\Lambda \in \mathscr{L}(E, F)$ such that

$$\begin{cases} \forall \epsilon > 0, \exists \eta_\epsilon > 0, \forall x, x_1 \in A, \\ (\|x - a\| \le \eta_\epsilon, \|x_1 - a\| \le \eta_\epsilon) \Longrightarrow \|\phi(x_1) - \phi(x) - \lambda.(x_1 - x)\| \le \epsilon \|x_1 - x\|. \end{cases}$$

When ϕ is strictly differentiable at a, then ϕ is Fréchet differentiable at a and $\Lambda = D\phi(a)$. We can find in [1] properties of the strict differentiability.

B.3.2 About Convexity

Definition B.3. When S is a convex subset of E (i.e., for all $x, x_1 \in S$, for all $t \in [0,1]$, $(1-t)x + tx_1 \in S$) and when $f : S \to \mathbb{R}$ is a concave function (i.e., for all $x, x_1 \in S$, for all $t \in [0,1]$, $f((1-t)x + tx_1) \ge (1-t)f(x) + tf(x_1)$), the subdifferential of f at $a \in S$ is

$$\partial f(a) := \{p \in E^* : \forall x \in S, f(a) - f(x) \ge \langle p, x - a \rangle\}.$$

We can find a complete theory of the subdifferential in [4, 79], for instance. Generally the authors treat the notion for a convex function; it suffices to multiply by (-1) a concave function to obtain a convex function, and to translate the results on convex functions into results on concave functions.

When a concave function f is Fréchet differentiable at a, then $\partial f(a) = \{Df(a)\}$, and we have the following important formula:

$$\forall x \in S, \quad f(a) - f(x) \ge \langle Df(a), x - a \rangle. \tag{B.17}$$

When S is an open convex set and f is concave, we define $\mathrm{Dif}(f)$ as the set of all $x \in S$ such that $Df(x)$ exists. Then $\mathrm{Dif}(f)$ is dense in S and $S \setminus \mathrm{Dif}(f)$ is Lebesgue-negligible [79] (Theorem 25.5, p. 246), and moreover the mapping $x \mapsto Df(x)$ is continuous on $\mathrm{Dif}(f)$. And so the concave functions appear to be "almost everywhere" continuously differentiable.

Definition B.4. When $a \in S$ and when S is convex, the normal cone of S at a is

$$N_S(a) := \{p \in E^* : \forall x \in S, \langle p, x - a \rangle \le 0\}.$$

B.3.3 Clarke Calculus

Following a famous theorem of Rademacher [67, 85] when A is open in E, when a function $f : A \to \mathbb{R}$ is locally Lipschitzian (i.e., for all $x \in A$, there exist a neighborhood N_x of x and a constant $c_x \in [0, +\infty)$ such that, for all $x_1, x_2 \in N_x$, $\|f(x_1) - f(x_2)\| \le c_x.\|x_1 - x_2\|$), f is Fréchet differentiable at each point of A except at most on a Lebesgue-negligible subset of A, and moreover the mapping

$x \mapsto Df(x)$ is continuous on $\mathrm{Dif}(f)$. And so the locally Lipschitzian functions appear as a generalization of the continuously differentiable functions.

Definition B.5. Let A be a nonempty open subset of E, $f : A \to \mathbb{R}$ be a locally Lipschitzian function, and $a \in A$. The Clarke differential of f at a is the subset $\partial f(a)$ defined as the convex hull of the set of all $\lim_{k \to +\infty} Df(x_k)$ when $(x_k)_{k \in \mathbb{N}} \in \mathrm{Dif}(f)^{\mathbb{N}}$ and when $\lim_{k \to +\infty} Df(x_k)$ exist in E^*.

We can find in [38] a treatment of this notion. We use the same notation for the subdifferential of a concave function and for the Clarke differential since, when f is concave, f is locally Lipschitzian [79] and the subdifferential of f at a coincides with the Clarke differential of f at a. When f is strictly differentiable at a, then f is Lipschitzian on a neighborhood of a [38] and $\partial f(a) = \{Df(a)\}$.

When $E = E_1 \times E_2$ is a product of normed spaces, then, when they exist, the partial Clarke differentials of f at $a = (a_1, a_2)$, denoted by $\partial_1 f(a)$ and $\partial_2 f(a)$, are the Clarke differentials of the partial functions $x_1 \mapsto f(x_1, a_2)$ and $x_2 \mapsto f(a_1, x_2)$.

Definition B.6. A function $f : A \to \mathbb{R}$ is called regular at the point a when the two following conditions are fulfilled:

(i) For all $v \in E$, $D^+ f(a; v) := \lim_{t \to 0+} \frac{1}{t}(f(a + tv) - f(a))$ exists in $\mathbb{R}$.

(ii) For all $v \in E$, $D^+ f(a; v) = \limsup_{y \to x, t \to 0+} \frac{1}{t}(f(y + tv) - f(y))$.

Theorem B.5. *If $(a_1, a_2) \in A \subset E_1 \times E_2$, and if the function $f : A \to \mathbb{R}$ is regular and Lipschitzian on a neighborhood of (a_1, a_2), then we have $\partial f(a_1, a_2) \subset \partial_1 f(a_1, a_2) \times \partial_2 f(a_1, a_2)$.*

This theorem is proven in [38] p. 48 (Proposition 2.3.15).

Theorem B.6. *Let $a \in A \subset E$, and consider two functions $f_1 : A \to \mathbb{R}$ and $f_2 : A \to \mathbb{R}$, which are Lipschitzian on a neighborhood of a, and two real numbers r_1, r_2. Then we have $\partial(r_1 f_1 + r_2 f_2)(a) \subset r_1 \partial f_1(a) + r_2 \partial f_2(a)$. Moreover when one of the two functions is strictly differentiable, then we have an equality instead of an inclusion.*

This theorem is proven in [38] p. 38–39 (Proposition 2.3.3, Corollary 1 and Corollary 2).

Following a classical idea in Mechanics and in Differential Geometry (probably due to Newton), if S is a smooth submanifold of E, a vector $v \in E$ is called a *tangent vector* to S at the point $a \in S$ when there exists a function $\varphi : (-\epsilon, \epsilon) \to S$, with $\epsilon \in (0, +\infty)$, which is differentiable at 0 such that $\varphi(0) = a$ and $\varphi'(0) = v$. When S is a smooth submanifold with boundary, we can define an *inward* vector to S at $a \in S$ as a vector of E for which there exists a function $\psi : [0, \epsilon) \to S$, with $\epsilon \in (0, +\infty)$, which is right-hand differentiable at 0 such $\psi(0) = a$ and $\psi'(0) = v$, i.e., $\lim_{t \to 0+} \frac{1}{t}(\psi(t) - a) = v$. If we discretize the time t in this last relation, we obtain the following notion.

Definition B.7. Let $a \in S \subset E$. The contingent cone of S at a is the set of all vectors $v \in E$ for which there exist $(x_k)_{k\in\mathbb{N}} \in S^{\mathbb{N}}$ and $(t_k)_{k\in\mathbb{N}} \in (0,+\infty)^{\mathbb{N}}$ such that $\lim_{k\to+\infty} t_k = 0$ and such that $v = \lim_{k\to+\infty} \frac{1}{t_k}(x_k - a)$.

We denote by $T_S^B(a)$ this contingent cone of S at a.

It is usual in the optimization theory to call "tangent" the "inward" vectors. And in the previous notation, the letter T stands for *tangent*, and B stands for Bouligand who is the creator of this contingent cone in 1932; see the references inside [3]. We can find other presentations of the contingent cone in [3, 38, 67].

Definition B.8. Let $a \in S \subset E$. The Clarke tangent cone of S at a is the set of all vectors $v \in E$ which satisfy the following property: for all $(z_k)_{k\in\mathbb{N}} \in S^{\mathbb{N}}$ which converges to a, for all $(t_k)_{k\in\mathbb{N}} \in (0,+\infty)^{\mathbb{N}}$ which is decreasing and which converges to 0, there exists $(y_k)_{k\in\mathbb{N}} \in S^{\mathbb{N}}$ such that $\lim_{k\to+\infty} \frac{1}{t_k}(y_k - z_k) = v$. The Clarke tangent cone of S at a is denoted by $T_S(a)$.

It is easy to see that the previous is equivalent to the characterization given in [38] (Theorem 2.4.5, p. 53).

Definition B.9. Let $a \in S \subset E$. S is called regular at a when $T_S(a) = T_S^B(a)$.

The previous definition is given in page 55 in [38].

Definition B.10. Let $a \in S \subset E$. The normal cone of S at a is

$$N_S(a) := \{p \in E^* : \forall v \in T_S(a), \langle p, v\rangle \leq 0\}.$$

Theorem B.7. *Let $(a_1, a_2) \in S_1 \times S_2 \subset E_1 \times E_2$ where E_1 and E_2 are two finite-dimensional real normed vectors spaces. Then we have $N_{S_1\times S_2}(a_1, a_2) = N_{S_1}(a_1) \times N_{S_2}(a_2)$.*

Theorem B.8. *Let $a \in S \subset E$. Then $N_S(a) = cl(\bigcup_{\lambda\in\mathbb{R}_+} \lambda \partial d_S(a))$, where cl means closure.*

This result is given in [38] p. 51 (Proposition 2.4.2).

B.3.4 A Multiplier Rule

We consider an open subset Ω in $\mathbb{R}^n$, real-valued functions g^0, $g^1, \ldots, g^{n_I}$, $h^1, \ldots, h^{n_E}$, and a nonempty subset S of Ω. With these elements we formulate the following maximization problem:

$$(\mathscr{P})\begin{cases}\text{Maximize } g^0(z) \\ \quad \text{when} \quad \forall \alpha \in \{1,\ldots,n_I\}, \;\; g^\alpha(z) \geq 0 \\ \qquad\qquad\quad \forall \beta \in \{1,\ldots,n_E\}, \;\; h^\beta(z) = 0 \\ \qquad\qquad\quad z \in S.\end{cases}$$

Note that the condition ($z \in S$) appears as an additional constraint when S is not open.

We set $d_S(x) := \inf\{\|x - y\| \; : \; y \in S\}$ the distance between x and S.

Definition B.11. The Clarke Lagrangian of ($\mathscr{P}$) is the function

$$L^C : \Omega \times \mathbb{R}^{1+n_I} \times \mathbb{R}^{n_E} \times \mathbb{R} \to \mathbb{R}$$

defined by

$$\begin{cases} L^C(z, \lambda_0, \lambda_1, \ldots, \lambda_{n_I}, \mu_1, \ldots, \mu_{n_E}, k) := \\ \displaystyle\sum_{\alpha=0}^{n_I} \lambda_\alpha g^\alpha(z) + \sum_{\beta=1}^{n_E} \mu_\beta h^\beta(z) - k.\|(\lambda, \mu)\|.d_S(z) \end{cases}$$

where $\lambda := (\lambda_0, \lambda_1, \ldots, \lambda_{n_I})$ and $\mu := (\mu_1, \ldots, \mu_{n_E})$.

Note that when $S = \Omega$ then $d_S(z) = 0$ for all $z \in \Omega$ and then $L^C(z, \lambda, \mu, k) = \mathscr{G}(z, \lambda, \mu)$ (given in Definition B.1) for all k.

The following result of Clarke comes from [38] (Theorem 6.11, p. 228) under the name "Lagrange Multiplier Rule."

Theorem B.9. *Let z_* be a solution of ($\mathscr{P}$). We assume that the functions g^0, g^1,...,g^{n_I}, h^1,...,h^{n_E} are Lipschitzian on a neighborhood of z_*. Then, for all sufficiently large $k \in R$, there exist real numbers λ_0, λ_1,..., λ_{n_I}, μ_1, ..., μ_{n_E} which satisfy the following conditions:*

(i) $\lambda_0, \lambda_1, \ldots, \lambda_{n_I}, \mu_1, \ldots, \mu_{n_E}$ *are not simultaneously equal to zero.*
(ii) *For all* $\alpha \in \{0, \ldots, n_I\}$, $\lambda_\alpha \geq 0$.
(iii) *For all* $\alpha \in \{1, \ldots, n_I\}$, $\lambda_\alpha g^\alpha(z_*) = 0$.
(iv) $0 \in \partial_1 L^C(z_*, \lambda_0, \lambda_1, \ldots, \lambda_{n_I}, \mu_1, \ldots, \mu_{n_E}, k) = 0$.

Using results of the Clarke calculus, the conclusion (iv) implies

$$0 \in \sum_{\alpha=0}^{n_I} \lambda_\alpha \partial g^\alpha(z_*) + \sum_{\beta=1}^{n_E} \mu_\beta \partial h^\beta(z_*) - k.\|(\lambda, \mu)\|.\partial d_S(z_*)$$

as it is explained in [38] (Remark 6.1.2, p. 228) which implies, by using Theorem B.8, that we have

$$0 \in \sum_{\alpha=0}^{n_I} \lambda_\alpha \partial g^\alpha(z_*) + \sum_{\beta=1}^{n_E} \mu_\beta \partial h^\beta(z_*) - N_S(z_*).$$

Note that when $S = \Omega$, the conclusion (iv) can be replaced by

$$0 \in \partial_1 \mathscr{G}(z_*, \lambda_0, \lambda_1, \ldots, \lambda_{n_I}, \mu_1, \ldots, \mu_{n_E}). \tag{B.18}$$

Remark B.4. When $S = \Omega$, problem $(\mathscr{P})$ is identical to problem $(\mathscr{M})$ of Sect. B.1. But the multiplier rule of Halkin is not a corollary of the multiplier rule of Clarke since a differentiable function at z_* is not necessarily locally Lipschitzian around z_*, and conversely the multiplier rule of Clarke is not a corollary of the multiplier rule of Halkin since a Lipschitzian function on a neighborhood of z_* is not necessarily Fréchet differentiable at z_*.

B.4 Karush–Kuhn–Tucker Theorems in Banach Spaces

B.4.1 Lagrange Principle

The following theorem is established in the book [1] (p.243) where it is called Lagrange principle.

Theorem B.10. *Let Ξ and Y be Banach spaces, and $\hat{\xi} \in \Xi$. We consider the following conditions:*

1. *$J : \Xi \to \mathbb{R}$ is a functional which is strictly differentiable at $\hat{\xi}$.*
2. *$F : \Xi \to Y$ is a mapping which is strictly differentiable at $\hat{\xi}$.*
3. *$Im(DF(\hat{\xi}))$ is closed into Y.*

If $\hat{\xi}$ is a solution of the following problem

$$\begin{cases} \text{Maximize} & J(\xi) \\ \text{when} & F(\xi) = 0 \\ & \xi \in \Xi \end{cases}$$

then there exist $\lambda_0 \in [0, +\infty)$ and $\Lambda \in Y^*$, a linear functional, such that the following conditions are satisfied:

(i) $(\lambda_0, \Lambda) \neq (0, 0)$.

(ii) $\lambda_0 DJ(\hat{\xi}) + \Lambda \circ DF(\hat{\xi}) = 0$.

Moreover, if $DF(\hat{\xi})$ is onto then we can choose $\lambda_0 \neq 0$.

B.4.2 Problems with Inequality Operator Constraints

The following theorem is established in the book [55] (p. 106, p. 111: Theorem 5.3, and p. 118: Theorem 5.6).

Theorem B.11. *Let Ξ, Y, and Z be three Banach spaces, and $\hat{\xi} \in \Xi$. We consider the following conditions:*

1. *Y is ordered by a cone C with a nonempty interior.*
2. *$\hat{S}$ is a convex subset of Ξ with a nonempty interior.*
3. *$\mathscr{F} : \Xi \to \mathbb{R}$ is a functional which is Fréchet differentiable at $\hat{\xi}$.*
4. *$g : \Xi \to Y$ is a mapping which is Fréchet differentiable at $\hat{\xi}$.*
5. *$h : \Xi \to Z$ is a mapping which is Fréchet differentiable at $\hat{\xi}$.*
6. *$S := \{\xi \in \hat{S} : g(\xi) \in -C, h(\xi) = 0\}$ is nonempty.*
7. *$Im(Dh(\hat{\xi}))$ is closed into Z.*

If $\hat{\xi}$ is a solution of the following minimization problem

$$\begin{cases} \text{Minimize } \mathscr{F}(\xi) \\ \quad \text{when} \quad \xi \in S \end{cases}$$

then there exist $\lambda_0 \in [0, +\infty)$, $\Lambda_1 \in Y^*$ a positive linear functional, $\Lambda_2 \in Z^*$ such that the following conditions are satisfied:

(i) $(\lambda_0, \Lambda_1, \Lambda_2) \neq (0, 0, 0)$.
(ii) $\langle \lambda_0 D\mathscr{F}(\hat{\xi}) + \Lambda_1 \circ Dg(\hat{\xi}) + \Lambda_2 \circ Dh(\hat{\xi}), \xi - \hat{\xi} \rangle \geq 0$ *for all* $\xi \in \hat{S}$.
(iii) $\langle \Lambda_1, g(\hat{\xi}) \rangle = 0$.

Moreover, if the following conditions are fulfilled

(Q1) $Dh(\hat{\xi})$ *is onto*
(Q2) *There exists* $\tilde{\xi} \in \text{int}\hat{S}$ *such that* $g(\hat{\xi}) + Dg(\hat{\xi})(\tilde{\xi} - \hat{\xi}) \in -\text{int}C$ *and* $Dh(\hat{\xi})(\tilde{\xi} - \hat{\xi}) = 0$

then we can choose $\lambda_0 \neq 0$.

References

1. V.M. Alexeev, V.M. Tihomirov, S.V. Fomin, *Commande Optimale*, French edn. (MIR, Moscow, 1982)
2. C.D. Aliprantis, K.C. Border, *Infinite Dimensional Analysis*, 2 edn. (Springer, Berlin, 1999)
3. J.-P. Aubin, *Viability Theory* (Birkhäuser, Boston, 1991)
4. J.-P. Aubin, *Mathematical Methods of Game and Economic Theory*, Revised edn. (Dover Publication, Inc., Mineola, New York, 2007)
5. J.-P. Aubin, I. Ekeland, *Applied Nonlinear Analysis* (Wiley Inc., New York, 1984)
6. T. Başar, G.J. Olsder, *Dynamic Noncooperative Game Theory*, 2 edn. (SIAM, Philadelphia, PA, 1999)
7. C. Berge, *Topological Spaces*, English edn. (Dover Publications, Inc., Mineola, New York, 1997)
8. J. Blot, *Infinite-Horizon Problems Under Holonomic Constraints*, Lecture Notes in Economics and Mathematical Systems, vol. 429 (Springer, Berlin, 1995) pp. 46–59
9. J. Blot, *Équation d'Euler-Lagrange*, in *Dictionnaire des sciences économiques*, ed. by C. Jessua, C. Labrousse, D. Vitry, D. Gaumont (Presses Universitaires de France, Paris, 2001) pp. 400–401
10. J. Blot, An infinite-horizon stochastic discrete-time Pontryagin principle, Nonlinear Anal.: Theor. Meth. Appl. **71**(12), e999–e1004 (2009)
11. J. Blot, Infinite-horizon Pontryagin principle without invertibility, J. Nonlinear Convex Anal. **10**(2), 157–176 (2009)
12. J. Blot, A Pontryagin principle for infinite-horizon problems under constraints, Dyn. Contin. Discr. Impul. Syst. Series B: Appl. Algor. **19**, 267–275 (2012)
13. J. Blot, *Infinite-Horizon Discrete-Time Pontryagin Principles via Results of Michel*, in *Proceedings of the Haifa Workshop on Optimisation and related topics*, ed. by S. Reich, A.J. Zaslavski, Contemporary Mathematics, vol. 568 (2012),0 pp. 41–51
14. J. Blot, P. Cartigny, Bounded solutions and oscillations of convex Lagrangian systems in presence of a discount rate, Zeitschrift für Analysis und ihre Anwendungen **14**(4), 731–750 (1995)
15. J. Blot, P. Cartigny, Optimality in infinite-horizon problems under signs conditions, J. Optim. Theor. Appl. **106**(2), 411–419 (2000)
16. J. Blot, H. Chebbi, Discrete time Pontryagin principle in infinite horizon, J. Math. Anal. Appl. **246**, 265–279 (2000)
17. J. Blot, B. Crettez, On the smoothness of optimal paths, Decis. Econ. Finance **27**, 1–34 (2004)
18. J. Blot, B. Crettez, On the smoothness of optimal paths II: some local turnpike results, Decis. Econ. Finance **30**(2), 137–180 (2007)

J. Blot and N. Hayek, *Infinite-Horizon Optimal Control in the Discrete-Time Framework*, SpringerBriefs in Optimization, DOI 10.1007/978-1-4614-9038-8,

19. J. Blot, B. D'Onofrio, R. Violi, Relative stability in concave Lagrangian systems, Int. J. Evol. Equat. **1**(2), 153–159 (2005)
20. J. Blot, N. Hayek, Second-order necessary conditions for the infinite-horizon variational problems, Math. Oper. Res. **21**(4), 979–990 (1996)
21. J. Blot, N. Hayek, Sufficient conditions for the infinite-horizon variational problems, Esaim-COCV **5**, 279–1292 (2000)
22. J. Blot, N. Hayek, Conjugate points in infinite-horizon optimal control problems, Automatica **37**, 523–526 (2001)
23. J. Blot, N. Hayek, *Infinite-Horizon Pontryagin Principles with Constraints* in Part 2 of *Communications of the Laufen Colloquium on Science*, ed. by A. Ruffing, A. Suhrer, J. Suhrer (Shaker Verlag, Aachen, 2007) 14 p. (reference Zentralblatt: Zbl 1134.49012)
24. J. Blot, N. Hayek, *Infinite horizon discrete time control problems for bounded processes*, Advances in Difference Equations, Volume 2008 (2008), Article ID 654267, 14 p. doi: 10.1155/2008/654267.
25. J. Blot, N. Hayek, F. Pekergin, N. Pekergin, The competition between Internet service qualities from a difference games viewpoint, Int. Game Theor Rev. **14**(1), (2012) 1250001 (36 pages), doi: 10.1142/SO 219198912500016.
26. J. Blot, N. Hayek, F. Pekergin, N. Pekergin, Pontryagin principles for bounded discrete-time processes, Optimization (2013) doi: 10.1080/02331.1934.2013.766991.
27. J. Blot, P. Michel, First-order necessary conditions for the infinite-horizon variational problems, J. Optim. Theor. Appl. **88**(2), 339–364 (1996)
28. J. Blot, P. Michel, The value-function of an infinite-horizon linear-quadratic problem, Appl. Math. Lett. **16**, 71–78 (2003)
29. J. Blot, P. Michel, On the Liapunov second method for difference equations, J. Difference Equat. Appl. **10**(1), 41–52 (2004)
30. V.G. Boltyanski, *Commande optimale des systèmes discrets*. French edn. (MIR, Moscow, 1976)
31. N. Bourbaki, *Éléments de mathématiques, topologie générale, Chapitres I à IV* (Hermann, Paris, 1971)
32. H. Brezis, *Functional Analysis, Sobolev Spaces and Partial Differential Equations* (Springer Science + Business Media LCC, New York, 2011)
33. W.A. Brock, On existence of weakly maximal programmes in a multi-sector economy, The Rev. Econ. Stud. **37**(2), 12970, 275–280.
34. D.A. Carlson, A.B. Haurie, A. Leizarowitz, *Infinite Horizon Optimal Control; Deterministic and Stochastic Systems*, 2nd, revised and enlarged edn. (Springer, Berlin, 1991)
35. G. Chichilnisky, An axiomatic approach of sustainable development, Soc. Choice Welf. **13**, 231–257 (1996)
36. G. Chichilnisky, P.J. Kalman, Application of functional analysis to models of efficient allocation of economic resources, J. Optim. Theor. Appl. **30**(1), 19–32 (1980)
37. C.W. Clark, *Mathematical Bioeconomics: Optimal Management of Renewable Resources* (Wiley Inc., Hoboken, NJ, 2005)
38. F.H. Clarke, *Optimization and Nonsmooth Analysis* (Wiley, New York, 1983)
39. G. Debreu, Valuation equilibrium and pareto optimum, Proc. Natl. Acad. Sci. USA **40**(7), 588–592 (1954)
40. K. Deimling, *Nonlinear Functional Analysis* (Springer, Berlin, 1985)
41. N. Dunford, J.T. Schwartz, *Linear Operators, Part I: General Theory* (Interscience Publishers, Inc., New York, 1958)
42. I. Ekeland, J. Scheinkman, Transversality conditions for some infinite horizon discrete time optimization problems, Math. Oper. Res. **11**, 216–229 (1986)
43. I. Ekeland, R. Temam, *Analyse Convexe et Problemes Variationnels* (Dunod and Gauthier-Villars, Paris, 1974)
44. S.D. Flam, R.J.B. Wets, Existence results and finite horizon approximates for infinite horizon optimization problems, Econometrica **55**(5), 1187–1209 (1987)
45. D. Gale, An optimal development in a multi-sector economy, The Rev. Econ. Stud. **34**(1), 1–18 (1967)

46. R. Grinold, Convex infinite horizon programs, Math. Program. **25**, 64–82 (1983)
47. H. Halkin, An abstract framework for the theory of process optimization, Bull. Amer. Soc. **725**, 677–678 (1966)
48. H. Halkin, Implicit function theorem and optimization problems without continuous differentiability of the data, SIAM J. Control **12**(2), 229–236 (1974)
49. H. Halkin, L.W. Neustadt, General necessary conditions for optimization problems, Proc. Nat. Acad. Sci. **56**(4), 1066–1071 (1966)
50. N. Hayek, A generalization of mixed problems with an application to multiobjective optimal control, J. Optim. Theor. Appl. **150**(3), 498–515 (2011)
51. N. Hayek, Infinite horizon multiobjective optimal control problem in the discrete time case, Optimization **60**(4), 509–529 (2011)
52. J.-B. Hiriart-Urruty, *Optimisation* (Presses Universitaires de France, Paris)
53. H. Hotelling, The economics of exhaustive resources, J. Polit. Econ. **39**(2), 137–175 (1931)
54. A. Ioffe, V.M. Tihomirov, *Theory of Extremal Problems* (North-Holland Publishing Company, Amsterdam, 1979)
55. J. Jahn, *Introduction to the Theory of Nonlinear Optimization*, 3rd edn. (Springer, Berlin, 2007)
56. L.V. Kantorovitch, G.P. Akilov, *Analyse fonctionnelle; tome 1*, French edn. (MIR, Moscow, 1981)
57. L.V. Kantorovitch, G.P. Akilov, *Analyse Fonctionnelle; tome 2: Équations Fonctionnelles*, French edn. (MIR, Moscow, 1981)
58. P.Q. Khanh, T.H. Nuong, On necessary optimality conditions in vector optimization problems J. Optim. Theor. Appl. **58** 63–81 (1988)
59. A.N. Kolmogorov, S.V. Fomin, *Éléments de la théorie de l'analyse fonctionnelle et de la théorie des fonctions*, French edn. (MIR, Moscow, 1977)
60. G. Köthe, *Topological Vector Spaces I*, English edn. second printing, revised (Springer, Berlin, 1983)
61. W. Krabs, S.W. Pickl, *Analysis, Controllability and Optimization of Time-Discrete Systems and Differential Games*, Lecture Notes in Economics and Mathematical Systems n° 529, (Springer, Berlin, 2003)
62. M.A. Krasnoselskii, *Positive Solutions of Operator Equations*, English edn. (P. Noordhoff Ltd., Groningen, 1964)
63. J. P. La Salle, *The Stability and the Control of Discrete Processes* (Springer, New York, 1986)
64. S. Lang, *Real and Functional Analysis*, 3rd edn. (Springer, New York, 1993)
65. C. Le Van, R.-A. Dana, *Dynamic Programming in Economics* (Kluwer Academic Publisher, 2002)
66. C. Le Van, H.C. Saglam, Optimal growth and the Lagrange multipliers, J. Math. Econ. **40**, 393–410 (2004)
67. C. Mayer, *Outils topologiques et métriques de l'ananlyse fonctionnelle; un cours de gustave choquet* (C.D.U. & S.E.D.E.S., Paris, 1969)
68. L.W. McKenzie, *Optimal Economic Growth, turnpike Theorems and Comparative Dynamics*, ed. by K.J. Arroxw, M.D. Intriligator. Handbook of Mathematical Economics, volume III , (North-Holland, Amsterdam, 1986) pp. 1281–1355
69. P. Michel, Une démonstration élémentaire du principe du maximum de Pontriaguine, Bull. Math. Économiques, **14**, 9–23 (1977)
70. P. Michel, Programmes mathématiques mixtes. Application au principe du maximum en temps discret dans le cas déterministe et dans le cas stochastique, RAIRO Recherche Opérationnelle, **14**(1), 1–19 (1980)
71. P. Michel, *Cours de mathématiques pour économistes*, 2nd edn. (Economica, Paris, 1989)
72. P. Michel, Some clarifications on the transversality conditions, Econometrica **58**(3), 705–728 (1990)
73. J. Milnor, Analytic proofs of the "hairy ball theorem" and the Brouwer fixed point theorem, Amer. Math. Monthly, **85**, 521–524 (1978)
74. H. Moulin, F. Fogelman-Soulié, *La convexité dans les mathématiques de la décision* (Hermann, Paris, 1979)

75. R. Pallu de la Barrière, *Cours d'automatique théorique* (Dunod, Paris, 19)
76. H.J. Pesch, M. Plail, *The Cold War and the maximum principle of optimal control*. Documenta Mathematica, Extra volume ISMP *optimization stories*, 331–343 (2012)
77. B.N. Pschenichnyi, *Necessary Conditions for an Extremum*, English edn. (M. Dekker, New York, 1971)
78. F.P. Ramsey, A mathematical theory of saving, The Econ. J. **38**(152), 543–559 (1928)
79. R.T. Rockafellar, *Convex Analysis* (Princeton University Press, Princeton, New Jersey, 1970)
80. P.A. Samuelson, *Les fondements de l'analyse économique; Tome 1: théorie de l'équilibre et principales fonctions économiques*, 2nd French edn. (Dunod, Paris, 1971)
81. L. Schwartz, *Topologie générale et analyse fonctionnelle* (Hermann, Paris, 1970)
82. A. Seierstad, K. Sydsaeter, Sufficient conditions in optimal control theory, Int. Econ. Rev. **18**(2), 367–391 (1977)
83. A.K. Skiba, Optimal growth with a convex-concave production function, Econometrica, **46**(3), 527–539 (1978)
84. E.D. Sontag, *Mathematical Control Theory* (Springer, New York, 1990)
85. E.M. Stein, *Singular Integrals and Differentiability Properties of Functions* (Princeton University Press, Princeton, New Jersey, 1970)
86. N.L. Stokey, R.E. Lucas, E.C. Prescott, *Recursive Methods in Economic Dynamics* (Harvard University Press, Cambridge, MA, 1989)
87. W.R.S. Sutherland, On optimal development in a multi-sectoral economy, Rev. Econ. Stud. **37**(4), 585–589 (1970)
88. M. Truchon, *Théorie de l'optimisation statique et différentiable* (Gaëtan Morin, Chicoutimi, Québec, 1987)
89. J. van Tiel, *Convex Analysis: An Introductory Text* (Wiley Chichester, 1984)
90. C.C. von Weizsäcker, Existence of optimal programs of accumulation for an infinite time horizon, Rev. Econ. Stud. **32**(2), 85–104 (1965)
91. M.L. Weitzman, Duality theory for infinite horizon convex models, Manag. Sci. **19**, 783–789 (1973)
92. J. Werner, *Optimization Theory and Applications* (Friedr. Vieweg & Sohn Verlagsgsllschaft mbH, Braunshweig, 1984)
93. A.J. Zaslavski, *Existence and Structure Results of Optimal Solutions of Variational Problems*, in *Recent Developments in Optimization and Nonlinear Analysis*, ed. by Y. Censor, S. Reich, Contemporary Mathematics, **204**, 247–278 (1997)
94. A.J. Zaslavski, *Turnpike Properties in the Calculus of Variations and Optimal Control* (Springer Science + Business Media, New York, NY, 2006)